Optics and Spectroscopy for Fluid Characterization

Optics and Spectroscopy for Fluid Characterization

Special Issue Editor

Johannes Kiefer

MDPI • Basel • Beijing • Wuhan • Barcelona • Belgrade

MDPI

Special Issue Editor
Johannes Kiefer
Universität Bremen
Germany

Editorial Office
MDPI
St. Alban-Anlage 66
Basel, Switzerland

This is a reprint of articles from the Special Issue published online in the open access journal *Applied Sciences* (ISSN 2076-3417) from 2017 to 2018 (available at: http://www.mdpi.com/journal/applsci/special_issues/optics_spectroscopy)

For citation purposes, cite each article independently as indicated on the article page online and as indicated below:

LastName, A.A.; LastName, B.B.; LastName, C.C. Article Title. *Journal Name* **Year**, *Article Number*, Page Range.

ISBN 978-3-03897-021-7 (Pbk)
ISBN 978-3-03897-022-4 (PDF)

Cover image courtesy of Johannes Kiefer.

Contents

About the Special Issue Editor

Johannes Kiefer, Prof. Dr.-Ing., is Chair Professor and Head of the division Technical Thermodynamics, at the University of Bremen, Germany. In addition, he is an Honorary Professor at the University of Aberdeen, Scotland, and he holds a guest professorship at the Erlangen Graduate School in Advanced Optical Technologies (SAOT) at the Friedrich-Alexander University (FAU) Erlangen-Nuremberg, Germany. He holds a degree in chemical engineering and a PhD from the FAU. During his postgraduate career, he worked at the FAU, the University of Lund, Sweden, and at the University of Aberdeen before he moved to Bremen in 2014. His research interests include developing and applying spectroscopic techniques for the characterization of advanced materials and processes.

applied sciences

MDPI

Editorial

Optics and Spectroscopy for Fluid Characterization

Johannes Kiefer [1,2,3,4]

[1] Technische Thermodynamik, Universität Bremen, Badgasteiner Str. 1, 28359 Bremen, Germany;
    jkiefer@uni-bremen.de; Tel.: +49-421-218-64777
[2] School of Engineering, University of Aberdeen, Aberdeen AB24 3UE, UK
[3] Erlangen Graduate School in Advanced Optical Technologies (SAOT), Friedrich-Alexander-Universität
    Erlangen-Nürnberg, 91052 Erlangen, Germany
[4] MAPEX Center for Materials and Processes, Universität Bremen, 28359 Bremen, Germany

Received: 15 May 2018; Accepted: 16 May 2018; Published: 21 May 2018

Abstract: This Editorial provides an introduction to and an overview of the special issue "Optics and Spectroscopy for Fluid Characterization".

Keywords: spectroscopy; tomography; holography; imaging; sensing; combustion; hydrogen bonding; process analytical technology; liquid

1. Introduction

All over the world, there is a huge and ever-increasing interest in the development and application of optical and spectroscopic techniques to characterize fluids in engineering and science. The large number of review articles that are frequently published in these areas is evidence of this. Recent examples have focused on applications of optical diagnostics to gas phase environments [1–5], liquids [1,4,6,7], and multiphase systems [7–10]. A key feature of such light-based methods is that they are usually non-intrusive, and hence they do not notably affect the system under investigation. As a consequence, optical techniques have been developed for many decades and represent the gold standard in many fields. The list of individual techniques utilizing absorption, refraction, diffraction and scattering effects is long and so is the list of the parameters that can be determined. The latter includes macroscopic properties such as temperature, chemical composition, thermophysical quantities, and flow velocity, but molecular information, e.g., about isomerism and intermolecular interactions, can also be obtained. This special issue entitled "Optics and Spectroscopy for Fluid Characterization" aims to demonstrate the breadth of the field in terms of methodology as well as applications.

2. Content of the Special Issue

The special issue starts with an educational and comprehensive review article [11] written by Andreas Fischer (University of Bremen) which lays out the basics of light scattering, highlighting its application to imaging flow velocimetry. The article describes the different flow measurement principles, as well as the fundamental physical measurement limits. Furthermore, the progress, challenges and perspectives for high-speed imaging flow velocimetry are considered.

The contributed articles discuss a large variety of methods and applications. Yan et al. [12] describe the use of a passive technique—flame emission spectroscopy—for in situ measurements of alkali metals evaporated during the incineration of municipal solid waste (MSW). They succeeded in detecting sodium, potassium, and rubidium species in the flame of an industrial incinerator. This suggests that their method is suitable for monitoring technical facilities for biomass and waste combustion. In another gas phase study, Shutov and colleagues [13] demonstrate a non-linear optical spectroscopic technique, termed Femtosecond Adaptive Spectroscopic Technique Coherent Anti-Stokes Raman Scattering (FAST CARS), to quantitatively map species' concentrations. Using the example of molecular oxygen, they illustrate how CARS can be used for the visualization of a gas flow in a free-space configuration. This method is proposed to be applicable to performing gas flow imaging utilizing any Raman-active species.

Liquid systems have been studied in a number of papers in this special issue as well. Collaborative activity by the Universities of Bremen and Manchester applied infrared spectroscopy to biosurfactants produced by microorganisms in a fermentation process [14]. Such biosurfactants represent amphiphilic compounds with polar and non-polar moieties and they can be used to stabilize emulsions, e.g., in the cosmetic and food sectors. They are highly viscous fluids and their structures may be affected when exposed to light and elevated temperatures. In this study, attenuated total reflection Fourier-transform infrared (ATR-FTIR) spectroscopy was applied to analyze the structure and aging of rhamnolipids as a representative of a complex biosurfactant. However, cell suspensions represent even more complicated systems that can be studied by optical techniques. Tang et al. [15] combined Raman spectroscopy with a discriminant analysis data evaluation approach to successfully distinguish between eight different cancerous human cells. The comparison of two methods—Linear Discriminant Analysis (LDA) and Quadratic Discriminant Analysis (QDA)—revealed better performance of the QDA when applied to the Raman spectra. Further, fluorescence imaging is another technique that is frequently used to study biological samples. The evaluation of such images, however, is not always straightforward. For this purpose, Zhu et al. [16] propose a sparse-representation-based image fusion method. They combine principle component analysis (PCA) to initially extract geometric similarities and classify the images. In a second step, the constructed dictionary is used to convert the image patches to sparse coefficients by a simultaneous orthogonal matching pursuit (SOMP) algorithm. As proof-of-concept, the proposed method is successfully applied to fluorescence images of biological samples.

The three remaining papers are concerned with fluid–solid interfaces that are widespread in nature, science, and engineering. Kiefer et al. [17] propose an infrared spectroscopic method to study metal–liquid interfaces that are of interest in electrochemistry and catalysis. They utilize an attenuated total reflection (ATR) spectroscopy approach in which a thin film of fluid is placed in between the ATR crystal and a metal plate. They obtain IR spectra from aqueous salt and acid solutions and an aluminum plate to demonstrate that useful information about the molecular interactions at the metal–liquid interface can be deduced. Chang et al. [18] employ a transmission infrared spectroscopy approach to study the molecular interactions between the ionic liquid 1-butyl-3-methylimidazolium trifluoromethanesulfonate and nano-sized alumina at elevated pressures. Interestingly, in contrast to the results obtained under ambient pressure, the local structures of both counter-ions appear disturbed under high pressure. They conclude that there is a formation of pressure-enhanced alumina/ionic liquid interactions under high pressure. Finally, liquid–solid interactions can also be used in a chemical sensing, in particular using surface-enhanced Raman scattering (SERS) spectroscopy, where the plasmonic enhancement of an electromagnetic field in the presence of nanostructured metal surfaces is utilized. Perozziello and co-workers [19] demonstrate that the development of new and efficient SERS substrates can be inspired by nature. They use natural nanomaterials with suitable structures and cover them with a thin gold layer. This approach allows high-sensitivity Raman spectroscopy to be performed at a relatively low cost, and thus it opens up new possibilities for the development of chemical and biochemical sensors.

3. Conclusions and Outlook

In conclusion, the papers in this special issue impressively demonstrate the huge diversity of the topic "Optics and Spectroscopy for Fluid Characterization". The ever-increasing availability of optical equipment in terms of light sources, detectors, and optical components at reasonable cost are important drivers of new developments in this field. In addition, a growing number of industries are realizing the potential of optical methods in terms of process analysis and material characterization. Therefore, it is foreseeable that the area of optics and spectroscopy for fluid characterization will experience further growth and will see fascinating new applications in the near and distant future.

Appl. Sci. **2018**, *8*, 828

Acknowledgments: The guest editor would like to thank all authors for submitting their excellent work to be considered for this special issue. Furthermore, he would like to thank all the reviewers for their outstanding job in evaluating the manuscripts and providing helpful comments and suggestions to the authors. The guest editor would like to thank the MDPI team involved in the preparation, editing, and managing of this special issue. This joint effort resulted in the above collection of high quality papers.

Conflicts of Interest: The authors declare no conflict of interest.

References

1.   Abram, C.; Fond, B.; Beyrau, F. Temperature measurement techniques for gas and liquid flows using thermographic phosphor tracer particles. *Prog. Energy Combust. Sci.* **2018**, *64*, 93–156. [CrossRef]
2.   Goldenstein, C.S.; Spearrin, R.M.; Jeffries, J.B.; Hanson, R.K. Infrared laser-absorption sensing for combustion gases. *Prog. Energy Combust. Sci.* **2017**, *60*, 132–176. [CrossRef]
3.   Cai, W.W.; Kaminski, C.F. Tomographic absorption spectroscopy for the study of gas dynamics and reactive flows. *Prog. Energy Combust. Sci.* **2017**, *59*, 1–31. [CrossRef]
4.   Kiefer, J. Recent advances in the characterization of gaseous and liquid fuels by vibrational spectroscopy. *Energies* **2015**, *8*, 3165–3197. [CrossRef]
5.   Ehn, A.; Zhu, J.; Li, X.; Kiefer, J. Advanced Laser-based Techniques for Gas-Phase Diagnostics in Combustion and Aerospace Engineering. *Appl. Spectrosc.* **2017**, *71*, 341–366. [CrossRef] [PubMed]
6.   Paschoal, V.H.; Faria, L.F.O.; Ribeiro, M.C.C. Vibrational Spectroscopy of Ionic Liquids. *Chem. Rev.* **2017**, *117*, 7053–7112. [CrossRef] [PubMed]
7.   Atkins, C.G.; Buckley, K.; Blades, M.W.; Turner, R.F.B. Raman Spectroscopy of Blood and Blood Components. *Appl. Spectrosc.* **2017**, *71*, 767–793. [CrossRef] [PubMed]
8.   Haven, J.J.; Junkers, T. Online Monitoring of Polymerizations: Current Status. *Eur. J. Org. Chem.* **2017**, *44*, 6474–6482. [CrossRef]
9.   Li, X.Y.; Yang, C.; Yang, S.F.; Li, G.Z. Fiber-Optical Sensors: Basics and Applications in Multiphase Reactors. *Sensors* **2012**, *12*, 12519–12544. [CrossRef]
10. Dinkel, R.; Peukert, W.; Braunschweig, B. In situ spectroscopy of ligand exchange reactions at the surface of colloidal gold and silver nanoparticles. *J. Phys. Condens. Matter* **2017**, *29*, 133002. [CrossRef] [PubMed]
11. Fischer, A. Imaging Flow Velocimetry with Laser Mie Scattering. *Appl. Sci.* **2017**, *7*, 1298. [CrossRef]
12. Yan, W.; Lou, C.; Cheng, Q.; Zhao, P.; Zhang, X. In Situ Measurement of Alkali Metals in an MSW Incinerator Using a Spontaneous Emission Spectrum. *Appl. Sci.* **2017**, *7*, 263. [CrossRef]
13. Shutov, A.; Pestov, D.; Altangerel, N.; Yi, Z.; Wang, X.; Sokolov, A.V.; Scully, M.O. Collinear FAST CARS for Chemical Mapping of Gases. *Appl. Sci.* **2017**, *7*, 705. [CrossRef]
14. Kiefer, J.; Radzuan, M.N.; Winterburn, J. Infrared Spectroscopy for Studying Structure and Aging Effects in Rhamnolipid Biosurfactants. *Appl. Sci.* **2017**, *7*, 533. [CrossRef]
15. Tang, M.; Xia, L.; Wei, D.; Yan, S.; Du, C.; Cui, H.-L. Distinguishing Different Cancerous Human Cells by Raman Spectroscopy Based on Discriminant Analysis Methods. *Appl. Sci.* **2017**, *7*, 900. [CrossRef]
16. Zhu, Z.; Qi, G.; Chai, Y.; Li, P. A Geometric Dictionary Learning Based Approach for Fluorescence Spectroscopy Image Fusion. *Appl. Sci.* **2017**, *7*, 161. [CrossRef]
17. Kiefer, J.; Zetterberg, J.; Ehn, A.; Evertsson, J.; Harlow, G.; Lundgren, E. Infrared Spectroscopy as Molecular Probe of the Macroscopic Metal-Liquid Interface. *Appl. Sci.* **2017**, *7*, 1229. [CrossRef]
18. Chang, H.-C.; Wang, T.-H.; Burba, C.M. Probing Structures of Interfacial 1-Butyl-3-Methylimidazolium Trifluoromethanesulfonate Ionic Liquid on Nano-Aluminum Oxide Surfaces Using High-Pressure Infrared Spectroscopy. *Appl. Sci.* **2017**, *7*, 855. [CrossRef]
19. Perozziello, G.; Candeloro, P.; Coluccio, M.L.; Das, G.; Rocca, L.; Pullano, S.A.; Fiorillo, A.S.; De Stefano, M.; Di Fabrizio, E. Nature Inspired Plasmonic Structures: Influence of Structural Characteristics on Sensing Capability. *Appl. Sci.* **2018**, *8*, 668. [CrossRef]

applied
sciences

MDPI

Review

Imaging Flow Velocimetry with Laser Mie Scattering

Andreas Fischer

Bremen Institute for Metrology, Automation and Quality Science (BIMAQ), University of Bremen, Linzer Str. 13, 28359 Bremen, Germany; andreas.fischer@bimaq.de; Tel.: +49-421-64600

Received: 9 November 2017; Accepted: 7 December 2017; Published: 13 December 2017

Abstract: Imaging flow velocity measurements are essential for the investigation of unsteady complex flow phenomena, e.g., in turbomachines, injectors and combustors. The direct optical measurement on fluid molecules is possible with laser Rayleigh scattering and the Doppler effect. However, the small scattering cross-section results in a low signal to noise ratio, which hinders time-resolved measurements of the flow field. For this reason, the signal to noise ratio is increased by using laser Mie scattering on micrometer-sized particles that follow the flow with negligible slip. Finally, the ongoing development of powerful lasers and fast, sensitive cameras has boosted the performance of several imaging methods for flow velocimetry. The article describes the different flow measurement principles, as well as the fundamental physical measurement limits. Furthermore, the evolution to an imaging technique is outlined for each measurement principle by reviewing recent advances and applications. As a result, the progress, the challenges and the perspectives for high-speed imaging flow velocimetry are considered.

Keywords: flow field measurement; measurement techniques; measurement uncertainty; physical limit; high-speed imaging

1. Introduction

Understanding, designing and optimizing flows is crucial, e.g., for improving fuel injections [1], combustions [2,3], fuel cells [4], wind turbines [5], turbomachines [6], human air and blood flows in medicine [7,8], as well as for the fundamental task of modeling flow turbulence [9]. For this purpose, optical measurement methods are essential tools that promise fast and precise field measurements of complex flows.

The advances of optoelectronic components and systems, in particular concerning powerful pulsed lasers (up to 1 J/pulse) with up to megahertz repetition rate and high-speed cameras with megapixel image resolution, allow qualitative flow imaging with ultra-high speed, up to 1 MHz [10]. The combination of powerful lasers and high-speed cameras enables the fast imaging of two or three flow dimensions. Note, however, that at the maximum megahertz acquisition rate, the available number of successive laser pulses is limited to about 100 pulses. Besides a qualitative understanding of complex flow phenomena, the flow field needs to be quantified. An essential physical quantity for understanding the flow behavior is the flow velocity, which is the sole measurand that is considered in the subsequent article.

In order to optically measure the flow velocity, the flow of interest is illuminated. For this purpose, a laser light source is usually applied, because laser light sources provide a high light power and a narrow linewidth. Both are beneficial for reducing the cross-talk between the optical measurement and the ambient light. The incident light is scattered on the fluid molecules or atoms, and the scattered light is detected and analyzed. As the size of the fluid atoms or molecules in the order of 0.1 nm is significantly smaller than the wavelength of visible light (0.4 μm–0.8 μm), the light scattering is so-called Rayleigh scattering. Optical flow velocity measurements based on Rayleigh scattering are described for instance in [11–17]. An overview article exists from Miles, 2001 [18]. The flow velocity measurements are based on the optical Doppler effect, which causes a shift in the observed scattered light frequency due to the scattering at the moving fluid particles. The evaluated Doppler frequency is proportional to the flow velocity in the inertial frame of the measurement system.

Rayleigh scattering has a comparatively low scattering efficiency, which means that the ratio between the scattered light intensity and the incident is low. For this reason, a temporal averaging in the order of 16 min is required to achieve an acceptable measurement uncertainty of 2% for the flow velocity images of a tube flow with a maximum velocity of 120 m/s [19,20]. Faster flow velocity measurements are possible for single point measurements. A single point measurement system with 32 kHz [21] was reported, as well as a 10 kHz measurement system with a minimal accuracy of 1.23 m/s [22]. Hence, the fast measurement of flow velocity images with kilohertz rates and a low uncertainty is only achievable by increasing the scattering efficiency of the scattering particles.

Increasing the scattering efficiency can be achieved by inserting scattering particles (if not naturally present), which follow the flow with negligible slip and do not disturb the flow. The insertion of particles is called seeding, and the particles are seeding particles. In the Rayleigh scattering regime, the scattering efficiency is proportional to the particle diameter to the power of four. Hence, doubling the particle diameter results in a 16-times higher scattering efficiency. This condition changes in the Mie scattering regime [23–25], i.e., when the particle size is near the wavelength of the light. As a compromise between a high scattering efficiency and a negligible slip, typical seeding particles have a size between 100 nm and several micrometer [26]. Hence, the use of seeding particles leads to optical flow measurements based on Mie scattering. While Rayleigh scattering is an elastic light scattering, which allows one to measure the velocity, the pressure and the temperature of the fluid [19,20], Mie scattering is an elastic light scattering that only allows one to measure the particle velocity. However, the scattering efficiency is increased by several orders of magnitude, which enables the measurement of flow velocity images with a lower measurement uncertainty at an identical spatial and temporal resolution with the same laser power. The other way around, the higher scattering intensity also allows a higher spatiotemporal resolution, image size and measurement rate at an identical measurement uncertainty with the same laser power. For this reason, the article is focused on optical flow velocity measurement techniques based on Mie scattering, which typically require seeding.

A large variety of different measurement techniques based on Mie scattering exists. In particular concerning the improved scattered light energy, one fundamental question is the existence and the magnitude of physical measurement limits, which hold for any measurement technique. Understanding these limits allows one to identify the potential for the future development of each measurement technique. Furthermore, the existence of a possibly superior technique needs to be clarified. Such a unified broad consideration of the different flow velocity measurement techniques based on Mie scattering, starting with a physical categorization of the different techniques, is missing.

For this reason, the aim of the article is to summarize and to review the developments, recent advances and fundamental limits of the different imaging flow velocity measurement principles based on Mie scattering. At first, the physical measurement principles are explained in Section 2. Next, the development towards an imaging technique is described for each principle in Section 3. Thereby, it is focused on the development of high-speed imaging measurement systems. Research results regarding fundamental limits of the measurement techniques are then presented in Section 4. Finally, application examples, as well as the respective challenges and perspectives are discussed in Section 5. Note that the focus is set on measurement techniques for meso-scale flow applications with dimensions between millimeter and meter. A current review of micro-scale flow metrology can be found in [27]. However, the present review is not regarding certain flow applications, but the development of the different flow measurement techniques and their fundamental measurement limits.

2. Measurement Approaches

2.1. Application of Mie Scattering

In order to obtain Mie scattering, the measurement approaches require scattering particles in the flow to be measured. If no natural scattering particles exist, artificial particles are added to the flow, which is known as seeding. The generation, the characterization and the application of different seeding particles is described in [26]. As an example, a typical liquid seeding material for flows with ambient temperature is diethylhexyl sebacate (DEHS), and a typical solid seeding material for flame flows is titanium dioxide (TiO_2). Due to the typically necessary seeding, the flow measurement approaches based on Mie scattering are intrusive techniques.

Since the light that is scattered on the particles is detected and evaluated, the velocity of the particle motion is measured. Note that it is a common case that the desired quantity is not the particle velocity. However, the particle velocity equals the flow velocity, if the particles follow the flow with no slip, if the particles have no self-motion and if the inserted particles do not change the flow behavior. For the derivation of the different measurement approaches, these ideal conditions are assumed to be fulfilled. The otherwise resulting limits of measurability are treated in Section 4.

The drawbacks of using seeding particles to achieve Mie scattering are accepted due to the advantage of a significantly increased scattered light intensity in comparison with flow velocity measurements based on Rayleigh scattering with no seeding particles. In order to quantify the improvement of the scattered light intensity, the calculated scattering cross-section is shown in Figure 1 as a function of the radius of a scattering particle that is assumed to be spherical and made of DEHS. The calculation is conducted according to [25] for a refractive index of 1.45, which is valid for DEHS at a light wavelength of 650 nm. In the Rayleigh scattering regime, the scattering cross-section increases proportional to the sixth power of the particle radius. For an increased scattering particle size from 0.1 nm (order of magnitude of a molecule, Rayleigh scattering) to 1 µm (typical order of magnitude of a seeding particle, Mie scattering), the scattering cross-section is increased by more than 16 orders of magnitude. Considering a spatial resolution of 100 µm^3 with air at normal pressure and room temperature as fluid, about 2×10^{13} fluid molecules are present according to the ideal gas law. Hence, the advantage of the higher scattering cross-section of seeding particles is partially equalized by the low number of scatterers. As a result, an increase of the scattering power of 1000 remains for the considered example. Another beneficial aspect of using seeding particles is the reduced light extinction, because the seeding is usually applied locally. Furthermore, the angular-dependent scattering and polarization effects have to be taken into account. For instance, Mie scattering has in general a stronger forward scattering than sidewards and backward scattering, which means a reduced light extinction in comparison to Rayleigh scattering. As a result, the potential of Mie scattering approaches for imaging flow measurements with acceptable uncertainty also at high measurement rates is illustrated, in particular with respect to the necessary distribution of the available light energy over space and time.

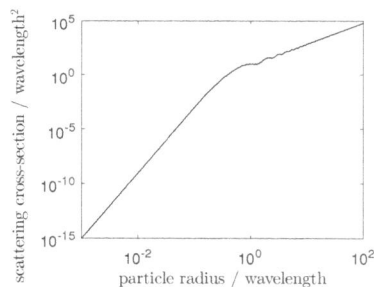

Figure 1. Calculated scattering cross-section over the particle radius normalized by the wavelength for a spherical particle made of diethylhexyl sebacate (DEHS) with a refractive index of 1.45 at 650 nm.

2.2. Scattered Light Evaluation

The velocity measurement approaches differ in the type of evaluation of the detected scattered light. One approach is to make use of the optical Doppler effect and to evaluate the momentum of the scattered light photons. Note that the photon momentum is Planck's constant divided by the wavelength, so that the photon momentum also represents the wavelength property or the energy of a photon. The second approach follows from the kinematic definition of the velocity and the evaluation of the position property of the scattered light photons. Hence, the developed measurement principles can be categorized into two groups [28]:

- Doppler principles,
- Time-of-flight principles.

The **Doppler principles** are based on the optical Doppler effect that occurs for light scattering at a moving object. In this fashion, the frequency shift (Doppler frequency) of the light scattered on a single particle (or multiple particles) is measured, and the Doppler frequency depends on the particle velocity. The relation between the particle velocity $\vec{v}_p$ and the Doppler frequency f_D reads: [29,30]

$$v_{oi} = \frac{\vec{o} - \vec{i}}{|\vec{o} - \vec{i}|} \cdot \vec{v}_p = \frac{\lambda}{|\vec{o} - \vec{i}|} f_D, \tag{1}$$

where λ denotes the wavelength of the incident light and v_{oi} is the measured velocity component along the bisecting line of the angle between the light incidence direction $\vec{i}$ from the illumination source and the observation direction $\vec{o}$ of the scattered light, cf. Figure 2. Note that the relation is an approximate solution for particle velocities significantly smaller than the light velocity, which applies for the flows considered here. As a result, a single velocity component is obtained with the given measurement configuration, and the vector $(\vec{o} - \vec{i})$ is the sensitivity vector. The measurement of all three velocity components requires three measurements with different incidence or different observation directions, so that the three sensitivity vectors span a three-dimensional space. Note further that Equation (1) is an approximate solution for velocities significantly smaller than the light velocity, which is applicable for the considered flows here. The remaining task is to determine the Doppler frequency of the scattered light signal, where the different measurement principles can be subdivided into Doppler principles with amplitude-based and frequency-based signal evaluation procedures.

The Doppler principles with an amplitude-based signal evaluation make use of the spectral transmission behavior of an optical filter. The optical filter converts the frequency information of the scattered light into an intensity information, which is finally measurable with a photodetector or camera. As the optical filter, an atomic or molecular filter or an interferometric filter is used. The interferometer can be a two-ray or a multiple ray interferometer. Note that the use of a filter requires laser light illumination with a significantly smaller linewidth than the linewidth of the filter transmission curve. Otherwise, the transmission curve of the filter is not resolved.

The Doppler principles with frequency-based signal evaluation also use interferometry. However, the scattered light is not superposed with itself, but with light with a different frequency. If the superposed light has no velocity-dependent shift in frequency, the measurement method is called a reference method. The superposed light is, e.g., from the illumination light source or a frequency comb. If the superposed light is also scattered light, but with a different Doppler frequency shift due to a different light incidence or observation direction, the difference of both Doppler frequencies is evaluated, and then, the measurement method is called the difference method.

light incidence
direction $\vec{i}$

observation
direction $\vec{o}$

v_{oi}

velocity $\vec{v}_p$
of the scattering particle

$\vec{o} - \vec{i}$

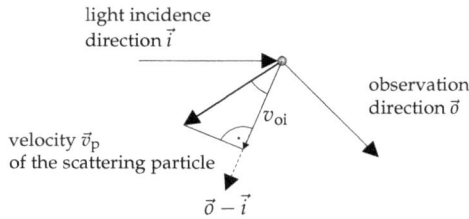

Figure 2. Measurement arrangement of Doppler principles illustrated for the light scattering on a single particle (measurement of one velocity component, acceleration neglected).

The **time-of-flight principles** are based on the kinematic velocity definition, which reads for one velocity component:

$$v_{p,x} = \dot{x} \approx \frac{\Delta x}{\Delta t}, \tag{2}$$

with Δx as the change in space during the time period Δt. The approximation in Equation (2) is valid when the particle acceleration during the measurement is negligible. This is a common assumption, which reduces the required number of pairs of position and time to two, cf. Figure 3. Either the change of the particle position for a given time period or the time period for a given spatial distance is then measured. Accordingly, the time-of-flight principles can be subdivided into two categories using time measurements or space measurements. Instead of two pairs of position and time, also a higher number of pairs can be acquired to take the acceleration into account where necessary [31–33]. Furthermore, the time-of-flight principles are also capable of performing three component measurements, e.g., with planar illumination, planar detection and at least two observation directions for triangulation.

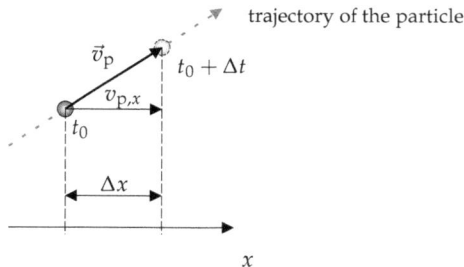

trajectory of the particle

$\vec{v}_p$

$t_0 + \Delta t$

$v_{p,x}$

t_0

Δx

x

Figure 3. Measurement arrangement of time-of-flight principles illustrated for the position measurement of a scattering particle (measurement of one velocity component, acceleration neglected).

An overview of the proposed categorization of the different measurement approaches is shown in Figure 4. Referring to the "16th International Symposium on Applications of Laser Techniques to Fluid Mechanics" in the year 2012, 79% of the 220 contributions contain or concern flow velocity measurements. An amount of 90% of these contributions is related to time-of-flight principles, while 10% is related to Doppler principles. For the articles concerning time-of-flight measurement principles, a majority of 99% considers or applies space measurement methods. This strong focus on space measuring time-of-flight principles is due to the commercial availability of respective measurement systems that are capable of simultaneously acquiring up to three components of three-dimensional velocity fields. As a result, these systems are widely applied, and the related research work is reported.

In contrast to this, measurement systems based on other principles currently seem to have a lag in development. However, in particular, Doppler principles are widely used as well and are advantageous for certain applications, which are illustrated for two examples. With a perpendicular arrangement of the illumination and observation direction, Doppler and time-of-flight principles allow one to measure different velocity components. For narrow optical accesses, which means low numerical apertures, the measurement of all three velocity components with a single principle is only achievable with additional observation or light incidence directions. If the additional optical access is not available or undesired, then the combination of both principles allows one to obtain measurements of all three velocity components with only one light incidence direction and one observation direction [34–36]. A further example of a beneficial combination of Doppler and time-of-flight principles is reported in [37], where a time-of-flight principle is used to resolve the velocity field image, but with a lower spatial and temporal resolution than the simultaneous single point Doppler measurement. As a result, turbulence investigations in an unsteady swirl flow could be performed together with an analysis of the global flow structures. The two examples show that hybrid approaches of combining Doppler and time-of-flight principles are possible and can offer advantages. The examples further demonstrate that each of the two fundamental measurement approaches has its own characteristics and benefits and that both approaches complement each other. In a physical sense, both approaches are complementary by evaluating the photon momentum or position.

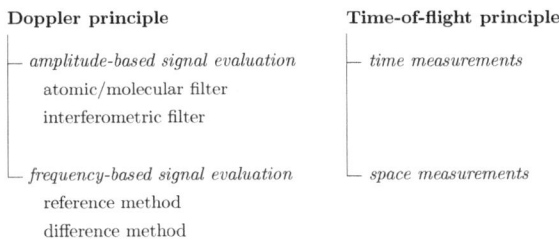

Doppler principle

— *amplitude-based signal evaluation*
    atomic/molecular filter
    interferometric filter

— *frequency-based signal evaluation*
    reference method
    difference method

Time-of-flight principle

— *time measurements*

— *space measurements*

Figure 4. Categorization of flow velocity measurement approaches based on Mie scattering.

3. Developed Measurement Techniques and Their Imaging Evolution

3.1. Fundamentals for the Evaluation of the Measurement Techniques

The quantity of interest is the fluid velocity (or flow velocity):

$$\vec{v}(\vec{x}, t), \tag{3}$$

which is in general a time-dependent vector field in the three-dimensional space with $\vec{x}$ as the space vector and t as the time variable. Hence, the flow measurements can be characterized and evaluated based on the following properties:

- The flow velocity is a vector quantity. Therefore, the number of measured velocity components is an important property. The abbreviated form 1c, 2c or 3c means that one, two or three components are measured, respectively.

- In order to characterize the spatial behavior of the flow velocity, the number of resolved space dimensions is an important property. Measurements are for instance pointwise, along a line, planar or volumetric, which is indicated by the abbreviated forms 0d, 1d, 2d or 3d, respectively. Characterizing the measurement along each space dimension is possible with the following parameters (cf. Figure 5a):

- – spatial resolution,
- – spatial distance between the adjacent measurements,
- – number of measurements along the respective space dimension or size of the measurement volume in the respective space dimension.

- In addition, the temporal behavior of the measured flow velocity is characterized with the parameters

- – temporal resolution,
- – temporal distance between the sequent measurements or measurement rate,
- – number of measurements along the time dimension or measurement duration.

The term measurement rate requires temporally-equidistant measurements. Otherwise, a mean measurement rate can be specified. The introduced terms are explained in Figure 5b. Note that the temporal resolution does not necessarily equal (or is smaller than) the reciprocal value of the measurement rate. Both quantities are independent.

- Each velocity value over space and time can finally be characterized with a measurement uncertainty by applying the international guide to the expression of uncertainty in measurements (GUM) [38,39]. According to the GUM, the measurement uncertainty is a parameter, associated with the result of a measurement, that characterizes the dispersion of the values that could reasonably be attributed to the measurand.

- Due to possible cross-sensitivities (for instance with respect to the temperature, vibrations or electromagnetic fields) or other impairments of the measurements (for instance limited optical access), it is important if the measurements concern non-reactive or reactive flows (flames), if the measurements are performed in a laboratory or in an industrial environment, if the measurement object is a simplified model or the real measurement object. For this reason, the surrounding and boundary conditions have to be described.

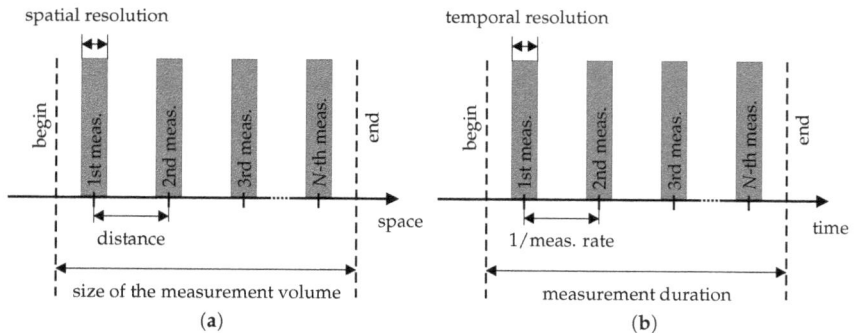

Figure 5. Illustration of the terms (**a**) spatial resolution, spatial distance of the measurements and size of the measurement volume along one space dimension and (**b**) temporal resolution, measurement rate and measurement duration.

In the following subsection, the development of each measurement approach from Section 2 to an imaging flow velocimetry system is summarized. Note that the summary is not meant to be exhaustive, but to give an overview of the state-of-the-art of current flow measurement systems. Since it is focused on the development of the respective measurement systems for achieving at least 2d1c measurements with high speeds, application examples, as well as the discussion of the measurement uncertainty are outsourced to Section 5.

3.2. Measurement Techniques Using the Doppler Principle

3.2.1. Amplitude-Based Signal Evaluation

In 1970, Jackson and Paul described a 0d reference laser Doppler velocimeter, where the scattered light with the Doppler frequency shift is analyzed with a confocal piezo-electrically-scanned Fabry–Pérot interferometer (FPI) [40]. The position of the maximal transmitted light signal through the FPI is a measure of the flow velocity. For 2d measurements, an FPI with planar mirrors can be used. Examples for such measurements are reported for Rayleigh scattering [12] and the fluid motion of the self-glowing Sun [41]. In 2013, Büttner et al. presented 1d flow measurements based on Mie scattering, but only for seven pixels [42]. The imaging to the FPI is performed with a self-made fiber-bundle with seven multi-mode fibers. For planar Doppler velocimetry with FPI (FPI-PDV), the crucial challenge is to achieve a high effective finesse of the interferometer. The finesse degrades with an increasing numerical aperture, which is required for maximizing the signal to noise ratio and for realizing an acceptable working distance. One solution to decouple the uncertainty between the FPI resolution and the numerical aperture was presented in 2003 by using single-mode fibers with an integrated FPI based on fiber Bragg grating technology [43]. A challenge for achieving a high signal to noise ratio is the low acceptance angle of single-mode fibers compared with multi-mode fibers. The fiber-integrated FPI approach was tested for 0d measurements on a rotating disc. The development of an FPI-PDV system for 2d flow measurements is an open task.

In place of a multiple ray interferometer, a two-ray interferometer can also be used, which is easier to apply for 2d measurements. In 1978, Smeets and George described a 0d reference laser Doppler velocimeter with a Michelson interferometer (MI) [44,45]. The extension to planar Doppler velocimetry with MI (MI-PDV) followed in 1982 by Oertel, Seiler and George [46,47], who named the measurement approach Doppler picture velocimetry (DPV). The mirror in the reference path of the interferometer is slightly tilted to generate a linear changing optical path difference and, thus, parallel interference fringes over the image. The velocity-dependent phase shift of these fringes is then determined with an image processing. The developed DPV system consists of a cwargon ion laser emitting at 514.5 nm and a charge coupled device (CCD) camera with 1374 × 1024 pixels and a shutter time of 2 ms [48]. An image series was not acquired, but CCD cameras usually provide a maximal frame rate of <60 Hz. In order to increase the sensitivity of the MI, a dispersive element (iodine vapor cell) was inserted in the reference path by Landolt and Rösgen in 2009 [49,50]. Furthermore, a normal CCD and an intensified CCD camera with an image resolution of 1008 × 1008 and 1280 × 1024 pixels are applied, respectively. The light sheets are generated with a pulsed Nd:YAG laser emitting at 532 nm with 100 mJ/pulse and a pulse width of 40 ns. The pulse repetition rate amounts to 10 Hz. High-speed flow imaging with MI-PDV remains to be investigated.

Almost in parallel, the usage of a Mach–Zehnder interferometer (MZI) was tested in 2007 by Lu et al. [51], which is an alternative to the MI. In addition, imaging fiber bundles are used to perform 2d3c measurements with a single CCD camera [52]. Each fiber bundle consists of 500 × 600 fibers and is 4 m long. An argon ion laser is used as the cw light source emitting at 514.5 nm. The available light power after the fiber that feeds the light sheet optics amounts to 200 mW. The capabilities of planar Doppler velocimetry with MZI (MZI-PDV) for time-resolved flow measurements with high-speed need further attention in future studies.

In contrast to interferometric filters, molecular or atomic filters are more robust with respect to mechanical vibrations. In addition, the transmission behavior does not depend on the light incidence direction, which simplifies the imaging. On the other hand, a suitable vapor material is required, which absorbs light at and near the laser wavelength. Observing the scattered light through the molecular filter, the transmitted light intensity is a measure of the Doppler frequency. The cross-sensitivity with respect to the scattered light intensity is corrected by applying a beam splitter and directly measuring the scattered light intensity. Hence, 2d measurements are achievable only by using two cameras for the photo detection, and each frame results in one flow velocity image.

The measurement technique is named Doppler global velocimetry (DGV), later also planar Doppler velocimetry (PDV). Komine described the technique first in 1990 in a patent [53]. There, he already mentioned resonance lines of atomic alkali metal vapors such as cesium (459 nm), rubidium (420 nm), potassium (404 nm), sodium (589 nm), lithium (671 nm), as well as molecular gases such as diatomic iodine and bromine with several resonance lines in the visible spectrum. Typically, the laser sources are combined with a molecular filter that contains iodine vapor. Komine further mentioned the pulsed frequency doubled Nd:YAQ laser (532 nm) and cw argon ion laser (514.4 nm) as possible light sources for instantaneous and time-averaged measurements, respectively. In 1991, such 2d3c flow velocity measurements were demonstrated with three observation units (each consisting of a camera pair, a beam splitter and a molecular filter) with different observation directions and a laser light sheet illumination [54]. Ainsworth et al. reviewed the progress in 1997 and described a method to reduce the required number of cameras by imaging both the signal image and the reference image onto a single camera [55]. Nobes introduced the fiber bundle technology in 2004 to measure all three velocity components on a single CCD camera with a single laser light sheet [56]. The fiber bundle consists of four arms each consisting of 500×600 fibers, which enables the scattered light detection from different observation directions with a single camera. As an alternative approach, Röhle and Willert presented in 2001 a 2d3c DGV system with a single observation unit and successively switching the light into one of the three light sheet optics with different illumination directions [57]. Charrett et al. extended this approach in 2014 for simultaneous measurements of all three velocity components by sinusoidally varying the intensity of each light sheet with a different frequency [58]. Analyzing the camera signals in the frequency domain allows one to separate the three channels. A performance test between two state-of-the-art DGV systems is described in [59]. With pulsed DGV systems, a temporal resolution of 10 ns is achieved. Due to the usage of CCD cameras and a low signal to noise ratio, the measurement rate of common DGV systems did not exceed a video rate of 25 fps. However, Thurow demonstrated in 2004 a measurement sequence of 28 images with 250 kHz [60]. The high repetition Nd:YAG laser system provided pulses with an average energy of 9 mJ/pulse at 532 nm, and the applied high-speed cameras consisted of at least 80×160 pixels.

Taking polarization effects into account and using a Faraday cell instead of a molecular absorption cell was considered by Bloom et al. in 1991 [61]. However, only 0d measurement systems have been developed with a Faraday cell to date [62].

In the classical DGV approach, two cameras (two measurements) are required to cope with two unknown quantities, the intensity and the frequency of the scattered light. In order to get rid of the second camera, Arnette et al. tried in 1998 the light sheet illumination with two different laser wavelengths and the simultaneous separate detection of the scattered light with a single polychrome camera [63]. However, the different scattering characteristics at two different wavelengths hinder measurements with polydisperse seeding particles. As a solution, the laser frequency can be modulated, and an image sequence is captured with a single camera. Note that the laser modulation and the image acquisition have to be faster than the movement of the seeding particles in the measurement volume, so that changes of the flow velocity and the scattered light intensity are negligible. This approach was developed by Charrett et al. in 2004 switching between two laser frequencies (2-ν-DGV) [64,65], by Müller et al. in 2004 switching between three laser frequencies also known as the frequency shift keying (FSK) technique (3-ν-FSK-DGV) [66,67], by Eggert et al. in 2005 switching between four different laser frequencies that allows a self-calibration (4-ν-FSK-DGV) [68,69], by Müller et al. in 1999 with a continuous sinusoidal laser frequency modulation (FM) and a signal evaluation in the frequency domain (FM-DGV) [70,71], which was developed further by Fischer et al. in 2007 [72–74], and by Cadel and Lowe in 2015 with a linear scan of the laser frequency and a detection of the position of the maximal transmitted scattered light through the molecular cell using a cross-correlation (CC) algorithm (CC-DGV) [75,76]. With the FM technique, a method for simultaneous 3c flow velocity measurements using three different illuminations with different incidence directions and different modulation frequencies was introduced by Fischer et al. in 2011 [77]. Later in 2015, the FM approach was combined with a light field camera to demonstrate simultaneous FM-DGV measurements at two parallel light sheets located at different depths [78]. For a topical model-based review of the different DGV variants with laser frequency modulation, which summarizes the laser sources and molecular filters and explains the different signal processing algorithms, we refer the reader to [79].

Except for FM-DGV, the developed measurement systems were not yet tested in combination with high-speed cameras, but with slower CCD cameras, which hinders continuous high-speed imaging of the flow velocity field. For FM-DGV, the developments started with a linear fiber array where each fiber is coupled with a photo detector resulting in 25 measurement channels. The applied modulation frequency of the laser frequency modulation amounts up to 100 kHz, and the measurement of steady flow velocity oscillations with frequencies up to 17 kHz was demonstrated [80,81]. The FM-DGV experiments with a high-speed camera based on complementary metal oxide semiconductor (CMOS) technology began in 2014 by Fischer et al. [82,83]. The developments culminated with a power amplified diode laser with fiber-coupling emitting 0.6 W at 895 nm, which is modulated with a maximal modulation frequency of 250 kHz. As a result, flow velocity images with 128×16 pixels and 128×64 pixels are resolved with a maximal measurement rate of 500 kHz and 250 kHz, respectively [84–86]. However, the effective bandwidth is usually smaller, because a sufficient signal to noise ratio is crucial to resolve unsteady flow phenomena with such high speeds.

As a result, the existing Doppler techniques with amplitude-based signal evaluation enable flow field measurements with a high data rate. The evaluation of superposed scattered light from multiple particles is possible with no additional signal processing, which is in contrast to Doppler techniques with frequency-based signal evaluation. However, the typically lower sensitivity and the cross-sensitivity with respect to scattered light intensity variations (over time or space) are critical issues. Comparing the usage of interferometric and atomic/molecular filters, atomic/molecular filters work with light absorption, while interferometric filters work with light reflection. Hence, atomic/molecular filters waste light that contains information. For this reason, the development of PDV systems with a two-ray interferometer as the interferometric filter and with high-speed cameras seems worth being studied in the future.

3.2.2. Frequency-Based Signal Evaluation

In 1964, only four years after the laser principle was demonstrated, Yeh and Cummings described a (reference) laser Doppler anemometer (R-LDA) with a He-Ne laser [87]. It is described as a reference method, i.e., the scattered light with the Doppler frequency shift is superposed with the non-shifted light, and the superposed light is measured with a photomultiplier. The frequency of the resulting beat signal is proportional to the desired flow velocity. In principle, the Doppler frequency can also be resolved with a laser frequency comb [88]. However, the capabilities of this approach for flow velocity measurements are an open question. The R-LDA approach is capable of 3c measurements when using three measurement systems with three different illumination directions. Due to the usage of a point detector, 0d measurements are possible.

Combining the LDA principle with the ranging principle known from radar allows one to perform 1d Doppler LiDAR measurements. The LDA principle was also extended to 2d by Coupland in 2000 [89] and also by Meier and Rösgen in 2009 [90]. The resulting technique is named heterodyne Doppler global velocimetry (HDGV). The major new component is a camera instead of point detector, which provides planar flow measurements. For this purpose, the flow is illuminated with a set of parallel laser beams or a light sheet, respectively. For instance, Meier and Rösgen used a cw laser (532 nm) with a power of 0.5 W and a smart pixel detector array with 144×90 pixels performing dual phase lock-in detection. The maximal tested image rate for a flow velocity field was 4.35 kHz, but only time-averaged results were presented.

An alternative approach to the reference LDA method is the difference LDA method by superposing the Doppler shifted light with light with a different Doppler frequency shift. The common measurement arrangement is the intersection of two laser beams, where the intersection region forms the measurement spot that allows 0d measurements. This technique is now known as (difference) laser Doppler anemometry/velocimetry (LDA/LDV), and an early review is presented in [91]. Further details regarding the LDA measurement technique and experimental setups can be found in the review [92] and the book [26]. The laser Doppler velocity profile sensor (LDV-PS) invented by Czarske in 2001 consists of two pairs of intersecting laser beams that allow one to resolve the 1d position of the crossing scattering particle within the intersection volume [93–95]. Note that the photo detection is still accomplished with a point detector, which hinders high data rates, which means multiple seeding particles in the intersection volume. In order to overcome this drawback and to yield 2d flow measurements, an imaging optic with a high-speed camera or a smart pixel detector array for the photo detection can be applied together with intersecting light sheets for the illumination. This technique was introduced as imaging laser Doppler velocimetry (ILDV) by Meier and Rösgen in 2012 [96]. The experimental setup is similar to the HDGV experiments. However, the maximal laser power was increased to 5.5 W. Using the high-speed camera, the beat signal at 512×64 pixels was directly acquired with 16 kfps. Again, the capabilities for time-resolved imaging measurements was estimated, but not yet tested in flow experiments.

To summarize, the development of imaging Doppler techniques with a frequency-based signal evaluation has been successful due to the progress of powerful lasers and high-speed cameras or smart pixel detector arrays. One key challenge is the increasing frequency of the beat signal with an increasing flow velocity, which currently limits the measurement range to about ± 1 m/s. Regarding this issue, the approach of a smart pixel detector array is promising, because it allows a fast, parallelized signal (pre-)processing on-chip. However, the necessity of acquiring multiple frames for one velocity image is a critical point.

3.3. Measurement Techniques Using the Time-Of-Flight Principle

3.3.1. Time Measurement

In 1968, Thompson realized a laser-based 0d1c flow measurement system by using two parallel laser beams with known distance and evaluating the transit time of the scattering particle that passes both beams [97,98]. This measurement technique is known as laser-2-focus anemometry (L2F). The approach was later extended for 2c and 3c measurements, but it remained a 0d measurement technique to date [34,99].

Instead of structuring the illumination, Ator proposed in 1963 a structured receiving aperture in 1963 [100]. The approach was validated for flow experiments by Gaster in 1964 [101]. When the image of the scattering particle crosses the parallel-slit reticle, the temporal distance between the two detected light pulses is directly proportional to the flow velocity. The technique is named spatial filter velocimetry (SFV). Reviews of the SFV development from 1987 and 2006 are contained in [102] and [103], respectively. As an example, Christoferi and Michel described in 1996 an SFV technique that avoids the blocking of light by directly using the columns of the camera pixels as the spatial filter [104]. This approach was later extended by employing the evaluation of quadrature signals [105]. In 2004, Bergeler and Krambeer proposed to use the orthogonal grating structure of the pixel matrix of a CMOS camera and to evaluate subsections of the image time series that enables 2d2c flow measurements [106]. A detailed system theoretic description of this SFV measurement technique is contained in [107]. Furthermore, tomographic SFV approaches for 3d3c flow measurements with multiple cameras were theoretically presented by Pau et al. in 2009 [108] and realized by Hosokawa et al. in 2013 [109].

Similar to the Doppler techniques with frequency-based signal evaluation, several image frames are necessary to obtain the flow velocity information. In addition, the frequency of the intensity modulation increases with an increasing flow velocity. For this reason, current SFV systems are typically restricted to flow velocities well below 1 m/s and flow imaging rates below the kHz range. However, the image resolution and the measurement rate are not well documented. Realizing a signal (pre-)processing directly on the camera chip as for instance was presented by Schaeper and Damaschke in 2011 [110] should be a subject for future research. Eventually, the SFV is an advantageous technique for imaging flow velocity fields with a high seeding particle concentration.

3.3.2. Space Measurement

The current standard field measurement technique is particle image velocimetry (PIV), where the seeded flow is illuminated by at least two consecutive light pulses and the light sheet is imaged with a camera. By cross-correlating sub-images (the so-called interrogation windows) that usually contain more than one particle and extracting the position of the correlation maximum, 2d2c flow velocity measurements are obtained. The name of PIV was introduced by Adrian in 1984 [111]. Reviews of the PIV development are available in [31,112,113].

Note that PIV requires short light pulses with a high energy per pulse and with a typical pulse width in the nanosecond range. Although light-emitting diodes can be used as the light source for PIV measurements, which was shown by Willert et al. in 2010 [114], shorter pulses with higher energy are achievable with pulsed lasers. This is in particular crucial for high-speed PIV measurements. Wernet and Opalsksi demonstrated in 2004 PIV measurements with a high-speed PIV system, which is capable of MHz measurement rates [35]. However, four CCD cameras are operated together with a delayed acquisition to finally obtain seven velocity vector maps out of eight frames. With two CMOS high-speed cameras each operated with 10 kHz or 25 kHz, Wernet demonstrated in 2007 continuous, time-resolved PIV measurements (tr-PIV) [115]. The cameras are capable of acquiring a time series of more than 10,000 frames each consisting of 1024×144 pixels or 640×80 pixels, respectively. For the light sheet generation, a Q-switched laser is operated at 10 kHz, which provides 6 mJ/pulse, and at 25 kHz with 2.5 mJ/pulse, respectively. The pulse length amounts to 130 ns.

With a light sheet illumination and a single camera, PIV enables 2d2c measurements. With a stereoscopic PIV setup, which means using two cameras with different viewing directions, 2d3c measurements are possible. Stereo-PIV was pioneered in 1991 by Arroyo and Greated [116]. High-speed stereo-PIV systems with 5 kHz and 10 kHz were developed for instance by Boxx et al., in 2009 and 2012, respectively [117–119]. A dual cavity, pulsed Nd:YAG laser emits at 532 nm pairs of light pulses with 2.6 mJ/pulse at repetition rates up to 10 kHz. The pulse duration is 14 ns. With a CMOS high-speed camera, 20 kfps are acquired with 512×512 pixels. The size of the interrogation window is 16×16 pixels.

Extensions of the stereo-PIV technique to 3d3c exist. The tomo-PIV technique is based on a volumetric illumination, multiple cameras, a tomographic reconstruction and the cross-correlation of interrogation volumes [120]. Another technique is holo-PIV, which is based on a holographic volumetric reconstruction and only requires one camera [121]. A high-speed system that combines the tomographic and holographic approaches was reported by Buchmann et al. in 2013 [122]. Furthermore, volumetric PIV measurements with a light-field camera form part of ongoing research [123–126]. A high-speed system based on light-field PIV is a topic for future research.

Note that standard PIV systems evaluate the particle positions in the images of two sequent light sheet pulses. It is mentioned for the sake of completeness that the accuracy in flows with strong acceleration can be increased by evaluating more than two image frames. The idea of such multi-frame PIV techniques goes back to Adrian in 1991 and was thoroughly investigated by Hain and Kähler in 2007 [31,32].

As an alternative to PIV, where image subsections are evaluated that usually contain several seeding particles, the technique to track the motion of each single particle, which is named particle tracking velocimetry (PTV), arose in 1991 [31,127]. Mass et al. demonstrated 3d3c measurements with multiple cameras in 1993 [128,129]. The 3d3c tracking with a single camera was introduced by Willert and Gharib in 1992 by using defocussing in connection with a coding aperture that is a mask with three pinholes [130]. As a result, the depth information is extracted from the particle image. In 1994 Stolz and Köhler proposed to discard the pinhole mask, to use monodisperse particles and to measure the diameter of the defocused particle image [131]. Cierpka et al., enhanced this technique for polydisperse seeding particles in 2010 to astigmatism-PTV (A-PTV) [132]. A cylinder lens is used in the imaging system, so that the particle depth information is coded in the ratio of the major axis of the elliptical particle images. Since processing each single particle is computationally intensive, Kreizer et al. proposed in 2010 a real-time image processing with a field programmable gate array (FPGA) [133,134]. Furthermore, PTV allows a higher spatial resolution in comparison to PIV, but the data rate is lower due to the requirement of a lower seeding particle concentration that enables a reliable pairing of particle images. In order to increase the possible seeding particle concentration to what is typically used for PIV, Cierpka et al. introduced a particle tracking based on more than two sequent images in 2013 [33]. In 2014, Buchmann et al. demonstrated high-speed measurements with A-PTV [135]. The illumination is provided by a pulsed light-emitting diode (LED) with a pulse repetition rate of 250 kHz and a pulse width of 1 μs. The imaging is performed with a single ultra-high-speed CCD camera with 312×260 pixels, which is capable of up to 1 Mfps and an exposure time of 250 ns.

As a result, PIV and PTV are well advanced measurement techniques for imaging flow velocities with high measurement rates. A comparison between the PIV and the DGV technique, which are the furthermost developed field measurement techniques based on the time-of-flight and the Doppler principle, respectively, is found in [136] with respect to their physical measurement capabilities. For instance, DGV is capable of coping with high seeding concentrations, and it is not necessary to resolve single particles or patterns of multiple particles in the image. Furthermore, a time-consuming image processing is not required for DGV, and each camera pixel can provide flow velocity information in each frame. Both features are essential for achieving high data rates. On the other hand, the laser requirements are lower for PIV, and the achievable accuracy is assumed to be higher. In order to understand the fundamental measurement capabilities, the fundamental measurement limits of the different flow measurement techniques are treated in Section 4.

4. Fundamental Measurement Limits

The aim of this section is to describe fundamental measurement limits, which apply for every flow velocimetry technique based on Mie scattering. The general effects of using seeding particles are discussed in Section 4.1. In the next Section 4.2, the measurement error due to the motion of the seeding particles not originating from the fluid motion is described. Finally, the physical limit of the measurement uncertainty due to photon shot noise is compared for Doppler and time-of-flight measurement principles in Section 4.3.

Note that each effect (discussed here, as well as others) propagates to a measurement uncertainty of the flow velocity. The computation of the uncertainty propagation and the combination of all uncertainty contributions is possible by means of analytic propagation calculations or Monte Carlo simulations, which is described in the international guide to the expression of uncertainty in measurements (GUM) [38,39]. For uncorrelated uncertainty contributions, the combined measurement uncertainty is the square root of the sum of all squared uncertainty contributions. As a result, the largest uncertainty contribution dominates and thus limits the total measurement uncertainty. However, the aim of the section is not to derive a complete measurement uncertainty budget, but to review fundamental measurement limits, which always occur in the measurement uncertainty analysis of flow velocimetry techniques based on Mie scattering.

4.1. Seeding

4.1.1. Influence on the Flow

In the case that no scattering particles are naturally contained in the flow, seeding particles need to be added. For instance, this always applies for dry air flows. The insertion of seeding particles influences the flow properties. Hence, the seeding as a part of the measurement system causes a retroaction to the flow, i.e., the measurement object.

Two typical seeding materials are listed in Table 1. DEHS particles are a typical choice for air flows at room temperature. The material is liquid at room temperature, so that the shape of small particles can be considered as approximately spherical. A common particle diameter is 1 µm. In hot flows such as in flames, particle materials with a higher boiling point are applied. As an example, solid particles of titanium dioxide (TiO_2) with a diameter of 0.4 µm can be applied.

Table 1. Material, density ρ_p and diameter d_p of two typical seeding particles.

Material	ρ_p (kg/m^3)	d_p (µm)
diethylhexyl sebacate (DEHS)	912	1
titanium dioxide (TiO_2)	3900	0.4

The retroaction of the seeding is difficult to quantify regarding the measurement uncertainty of the flow velocity. However, the influence on the Reynolds number Re, which is an important similarity figure for flow experiments, can be derived. The Reynolds number is obtained for the flow around or through a body with the characteristic length L, the fluid density ρ, the dynamic fluid viscosity η and the characteristic flow velocity v (relative to the body) [137]:

$$Re = \frac{\rho v L}{\eta}. \tag{4}$$

As a result, the seeding influences the mean fluid density and the fluid viscosity, which is the effect of the retroaction of the seeding.

An example is given in Figure 6a, where the change of the mean air density (at room temperature) is shown as a function of the seeding particle concentration considering for DEHS and TiO_2 (cf. Table 1) as the seeding material, respectively. Even at particle concentrations of 10^{13} /m^3, which are hardly achievable [138,139], the density change is below 1%. Since the particle concentration is typically several orders of magnitude lower, the influence of the seeding on the air flow density is usually negligibly small. The change in viscosity and the general change of the flow due to the presence of two phases is not known here. However, the low volume percentage of the seeding leads to the assumption that it is negligible as well. The volume percentage is shown in Figure 6b over the particle concentration. It is always below 0.001%. As a result, the retroaction of the seeding is usually negligible compared with other error sources.

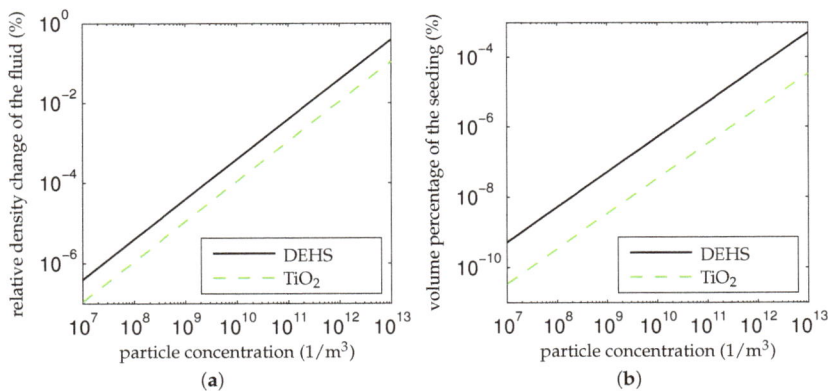

Figure 6. (a) Change of the mean density of the fluid due to the seeding and (b) volume percentage of the seeding over the particle concentration.

4.1.2. Flow Sampling Phenomenon

Since not the flow velocity, but the seeding particle velocity is actually measured, the flow field is sampled in space. The sampling is not equidistant due to the random position of the seeding particles. Hence, the determined particle velocities are a random sample of the flow velocity field to be measured. As a result of the flow sampling, the complete spatiotemporal flow behavior cannot be reconstructed from the velocities of the particles in general, even if the particle positions and the measurement time stamps are exactly known.

Therefore, it is desirable to increase the sample size, which is possible by increasing the seeding particle concentration and by decreasing the temporal and spatial resolution (spatiotemporal averaging). Increasing the seeding particle concentration is only possible to a certain extent, because this deteriorates the optical access due to the increased light extinction. In addition, except for DGV, single particles or at least particle patterns need to be resolved in the measurement volume.

Assuming a (locally) temporally-constant flow, the temporal resolution can be reduced by increasing the averaging time. One typical example is comprised of LDA measurements in stationary boundary layer flows [140]. Assuming a (locally) spatially-constant flow, the spatial resolution can be reduced by increasing the spatial averaging. Many flows exist, which are neither temporally nor spatially constant. An example is an eddy with an almost fixed structure during the measurement, which moves due to a non-zero mean flow velocity [141]. Due to the translational motion of the flow velocity field of the eddy, a temporal averaging also means a spatial averaging according to Taylor's hypothesis, and vice versa [142]. As a result, the resolution of flow structures can be limited by spatial or temporal averaging.

Actually, measurement systems always have a limited spatial and temporal resolution. For this reason and due to the flow sampling phenomenon, the measured mean particle velocity in general exhibits random fluctuations and is not identical to the true mean flow velocity, if spatial (eddies, gradients) or temporal (turbulence) fluctuations of the flow velocity occur. Space, time and velocity follow an uncertainty principle, which leads to a fundamental limit of measurability.

The finding is illustrated as an example. A stationary flow is considered with a velocity gradient in the measurement volume. The dimension of the measurement volume is the finite spatial resolution. The resulting measurement uncertainty is described by Durst et al. in [143–145]. Fischer et al. derived the relation between the spatial resolution, the temporal resolution and the resulting standard deviation σ_v of the measured flow velocity in the measurement volume [139]. Assuming a uniform distribution of the seeding particles in space, neglecting the particle transit time through the measurement volume in comparison with the temporal resolution Δt and neglecting the flow velocity variations in the measurement volume in comparison with the mean flow velocity $\bar{v}$, the relation reads:

$$\sigma_v \sim \frac{G}{\sqrt{c_p |\bar{v}| \Delta t}} \sqrt{\frac{\Delta x}{\Delta y}}. \tag{5}$$

The symbol G denotes the absolute value of the flow velocity gradient, c_p is the particle concentration and Δx, Δy are the respective spatial resolution. Both axes are perpendicular to the flow direction, and x is oriented parallel and y perpendicular to the direction of the flow velocity gradient. The proportionality constant of Equation (5) results from the light intensity distribution of the illumination in the measurement volume and can be determined with a simulation [139]. Hence, an uncertainty in space (finite spatial resolution) and a spatial flow velocity gradient leads to a flow velocity uncertainty. Similar to this finding, a velocity uncertainty can be derived from an uncertainty in time (finite temporal resolution) and a temporal velocity change.

Furthermore, Nobach [146] and Fischer et al. [79] studied the influence of fluctuations of the detected scattered light on PIV and DGV measurements, respectively. The scattered light fluctuations are a direct consequence of the discrete scattering particles moving through the measurement volume, but note that the light fluctuations also depend on the actual illumination and observation aperture of the measurement volume and the variation of the seeding particle concentration.

4.2. Particle Motion

4.2.1. Flow Following Behavior

Seeding-based measurement techniques rely on the assumption that the seeding particles follow the flow with no slip. Since this assumption is only approximately valid, the particle-fluid interaction leads to a fundamental measurement limit.

By neglecting the interaction between the particles and assuming a homogeneous, laminar flow field around the particle, Basset derived the equation of motion for a spherical particle with a homogeneous material (uniform composition) and a smooth surface [147]:

$$\frac{\pi}{6}d_p^3 \rho_p \frac{d\vec{v}_p}{dt} = -3\pi\eta d_p (\vec{v}_p - \vec{v}) + \frac{\pi}{6}d_p^3 \rho \frac{d\vec{v}}{dt} - \frac{1}{2}\frac{\pi}{6}d_p^3 \rho \frac{d(\vec{v}_p - \vec{v})}{dt}$$

$$- \frac{3}{2}d_p^2 \sqrt{\pi\rho\eta} \int_{t_0}^{t} \frac{d(\vec{v}_p - \vec{v})}{dt'} \frac{1}{\sqrt{t - t'}} dt' + \vec{F}. \tag{6}$$

The symbols $\vec{v}_p$, $\vec{v}$ are the particle and the flow velocity; ρ_p, ρ are the particle and the fluid density; η is the dynamic fluid viscosity; d_p the is particle diameter; t_0 is the initial time; and $\vec{F}$ represents the external forces acting on the particle, e.g., the centrifugal force in a vortex, the buoyancy force in a shear flow or the gravitational force [148–150]. External forces are not considered here. A detailed derivation of the equation of motion is contained for instance in [151]. The condition of no particle interaction is considered to be applicable, if the mean distance between the particles is larger than 1000 times the particle diameter [26]. For $d_p = 1\,\mu m$, this means for instance a maximal particle concentration of about $10^9\,/m^3$ [152]. For this reason, the condition of no particle interaction is usually applicable. In particular for DGV measurements, however, higher particle concentrations can occur [139], and particle interactions have to be taken into account. For the approximation of the relation between the mean free particle distance and the particle concentration, the space for each particle is considered as a cube with identical dimension. The edge length of the cube can be interpreted as the mean free particle distance. The resulting relation is shown in Figure 7 for two different particle diameter d_p.

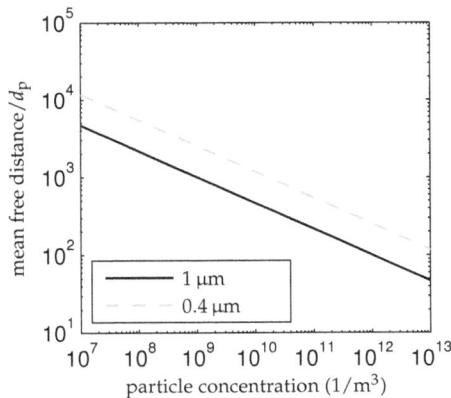

Figure 7. Mean free distance of the seeding particles divided by the particle diameter d_p over the seeding particle concentration for two different particle diameters.

For gaseous flows, it is usually $\rho_p \gg \rho$, so that Equation (6) can be simplified according to Hjemfelt and Mockros [152,153]. One component of the velocity then follows the relation:

$$\frac{dv_p}{dt} = -\underbrace{\frac{18\eta}{\rho_p d_p^2}}_{=1/\tau_p}(v_p - v) \tag{7}$$

with v_p, v as the components of the particle and flow velocity, respectively, and τ_p as the characteristic time constant of the particle. The deviation Δv between the particle and the flow velocity for a spherical particle thus reads:

$$\Delta v = v_p - v = -\tau_p \frac{dv_p}{dt}. \tag{8}$$

As an example, the time constant amounts to 3.4 μs and 2.3 μs for the particles listed in Table 1 in air flows at room temperature ($\eta = 15 \times 10^{-6}$ Pas, $\rho = 1.2\,\text{kg/m}^3$). Note that an alternative and well-known figure of merit for the measurement deviation is the slip [26]:

$$s = \frac{v - v_p}{v} = -\frac{\Delta v}{v}. \tag{9}$$

As a result, the measurement deviation Δv depends on the characteristic time constant and the acceleration of the particle, which includes spatial and temporal changes of the flow velocity. Since both quantities are usually not available during the flow measurement, the measurement deviation is a fundamental measurement limit. The limit is calculated as an example, assuming a cosinusoidal variation of the flow velocity with the frequency f, the amplitude v_a and the phase φ_a. Solving the equation of motion Equation (7) of the particle and inserting the solution into Equation (8) leads to the measurement deviation:

$$\Delta v = v_a \frac{f}{\sqrt{f^2 + f_g^2}} \cos\left(2\pi f t + \varphi_a - \arg\left(1 + j\frac{f}{f_g}\right) - \frac{\pi}{2}\right), \tag{10}$$

with j as imaginary unit and $f_g = 1/(2\pi\tau_p)$ as the 3 dB limit frequency of the transfer function:

$$\frac{\underline{V_p}(j\omega)}{\underline{V}(j\omega)} = \frac{1}{1 + j\frac{f}{f_g}}, \qquad \text{with } \omega = 2\pi f. \tag{11}$$

For a harmonic oscillation of the flow velocity with the frequency f, the measurement deviation of a single velocity value is in the interval $\pm v_a \frac{f}{\sqrt{f^2 + f_g^2}}$. Without knowledge of t, f, f_g and φ_a, this represents a fundamental measurement limit.

For the particles listed in Table 1 in air flows at room temperature, the 3 dB limit frequency is $f_g = 46.8$ kHz and $f_g = 69.2$ kHz, respectively. Considering $f = 10$ kHz for instance, the maximal deviation $|\Delta v|$ then amounts to $0.21 \times v_a$ and $0.14 \times v_a$. It is usually the case that the amplitude v_a and/or the phase φ_a of the velocity oscillation is of interest from a time series measurement of the flow velocity. Taking the particle-following behavior into account, the amplitude and phase deviations according to the Equation (11) read for a frequency f:

$$\Delta v_a(f) = v_a \cdot \left(\frac{1}{\sqrt{1 + \frac{f^2}{f_g^2}}} - 1 \right) \tag{12}$$

$$\Delta \varphi_a(f) = - \arg \left(1 + j \frac{f}{f_g} \right). \tag{13}$$

For the particles listed in Table 1, the resulting amplitude and phase deviation with respect to the frequency is shown in Figure 8. Comparing the amplitudes of the flow and particle velocity oscillation for instance, the relative measurement deviation for $f = 10\,\text{kHz}$ is -2.2% and -1%, respectively.

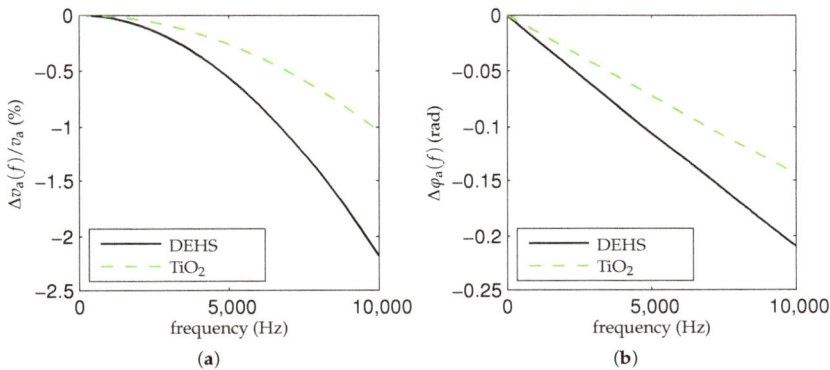

Figure 8. Deviation of (**a**) the amplitude and (**b**) the phase of a cosinusoidal excitation due to the particle-following behavior of the particles listed in Table 1.

If the oscillation frequency f and the characteristic time constant τ_p or the limit frequency f_g of the particle is known, a correction of the measurement deviation is possible in principle. However, this is typically not the case. Hence, it is shown for the example of a harmonic flow oscillation that the particle-following behavior leads to a fundamental measurement limit.

4.2.2. Brownian Motion

Due to the Brownian motion of the molecules, the particle velocity fluctuates and a random error results for the flow velocity measurement [154]. The two-sided noise power spectral density $S_v(f)$ of the velocity is [155,156]:

$$S_{v_p}(f) = \frac{2D}{1 + \frac{f^2}{f_g^2}}, \quad \text{with } D = \frac{k_B T}{3\pi \eta d_p} = \frac{k_B T \tau_p}{m_p}, \tag{14}$$

with D as diffusion constant, which is connected with the Boltzmann constant k_B, the temperature T, as well as the particle diameter d_p or the particle mass $m_p = \rho_p \frac{1}{6} \pi d_p^3 = 3\pi \eta d_p \tau_p$. According to Parseval's theorem, the maximal resulting standard deviation of the measured flow velocity reads:

$$\max(\sigma_v) = \sqrt{\int_{-\infty}^{+\infty} S_{v_p}(f) \mathrm{d}f} = \sqrt{\frac{D}{\tau_p}} = \sqrt{\frac{k_B T}{m_p}}. \tag{15}$$

For the particles listed in Table 1 in air flows at room temperature ($\eta = 15 \times 10^{-6}$ Pas, $\rho = 1.2 \, \text{kg/m}^3$), the uncertainty amounts to 3 mm/s and 6 mm/s, respectively. The order of magnitude of the uncertainty indicates that the Brownian particle motion is in particular of interest for the measurement of slow fluid motions or the usage of light particles. Actually, the resulting measurement uncertainty due to Brownian motion is even smaller, because the flow velocity measurement is averaged over a time period Δt and usually also over the velocity of N_p particles. For $\Delta t \gg 2\tau_p$ and approximating the spectral noise density as band limited white noise, the measurement uncertainty limit then is:

$$\sigma_v \approx \sqrt{\frac{2D}{N_p \Delta t}}. \tag{16}$$

Note that the number of particles results from the particle concentration and the spatial resolution of the measurement system. Hence, the Brownian motion of the particles leads to a fundamental uncertainty relation between flow velocity, space and time.

4.2.3. Illumination Effects

The light illumination influences the measurement due to the photon momentum (particle motion) and due to the light absorption (particle heating). Both effects are described by Durst et al. in [157,158]. Concerning the radiation pressure, the induced force on a spherical particle due to the photon momentum results in the maximal measurement error:

$$\Delta v = \frac{I \cdot d_p}{12 c \eta} \tag{17}$$

with c as light velocity, d_p as particle diameter, η as the dynamic viscosity of the flow fluid and I as the incident light intensity. For the seeding particles listed in Table 1 in air flows at room temperature ($\eta = 15 \times 10^{-6}$ Pas) and $I = 100 \, \text{W/mm}^2$, i.e., assuming a light power of 1 W distributed over a cross-sectional area of 100 µm times 100 µm, the measurement error is maximal 2 mm/s and 1 mm/s, respectively. For imaging techniques, the light intensity is typically smaller and the resulting error negligible.

Furthermore, multiple scattering on particles occurs, which can be crucial for Doppler measurement techniques. An introduction to this phenomenon regarding DGV measurements is given in [139].

4.3. Photon Shot Noise

Considering the typical illumination with narrow band laser light, the number of detected scattered photons exhibits a Poisson distribution [159]. This phenomenon is known as photon shot noise, which ultimately limits the achievable measurement uncertainty. As a characteristic behavior of shot noise limited measurements, the measurement uncertainty is indirectly proportional to the square root of the number of photons. Shot noise is an important fundamental measurement limit in particular for the measurement of flow velocity images with a high measurement rate, because the limited number of available photons or light energy, respectively, needs to be distributed over space and time.

The minimal achievable measurement uncertainty due to photon shot noise is obtained from the square root of the Cramér–Rao bound [160,161]. The Cramér–Rao bound is explained for instance in [162,163]. As an alternative approach, Heisenberg's uncertainty principle [164] can also be applied, which was recently demonstrated by Fischer for L2F and LDA [165]. In the following, LDA (difference method) and DGV (without laser frequency modulation) are considered as a representative Doppler technique frequency-based and amplitude-based signal evaluation, respectively. L2F and PTV are selected to represent time-of-flight techniques with time and space measurements, respectively.

For DGV, the first error propagation calculation including photon shot noise is from McKenzie in 1996 [138]. Fischer et al. investigated the Cramér–Rao bound due to photon shot noise for DGV with and without laser frequency modulation in 2008 and 2010 [73,166,167]. A comprehensive study of the shot noise limits of the different DGV techniques is contained in a topical contribution by Fischer [79].

Several studies covered the calculation of the Cramér–Rao bound for LDA with respect to additive white Gaussian noise with constant variance [168–171]. An error propagation calculation for noise with a Poisson distribution was given by Oliver in 1980 [172] and the calculation of the Cramér–Rao bound followed by Fischer in 2010 and 2016 [165,167].

Regarding L2F, an error propagation calculation for Poisson statistics was performed by Oliver in 1980 [172]. The calculation of the Cramér–Rao bound was demonstrated by Lading and Jørgensen in 1983 and 1990 [173,174]. A detailed derivation is also contained in [167] and [165].

The Cramér–Rao bound for the particle location measurement in PTV was investigated for photon shot noise by Wernet and Pline in 1993 [175]. The calculation was extended by Fischer in 2013 and adapted to evaluate the Cramér–Rao bound for the velocity measurement [176]. It is mentioned for the sake of completeness that similar PIV error considerations were performed under the assumption of white Gaussian noise by Westerweel in 1997 and 2000 [177,178]. Regarding photon shot noise, the result of the Cramér–Rao bound for PTV is applicable for PIV measurements with a single scattering particle.

The majority of the investigations concern a single measurement technique. A comparison of the minimal achievable uncertainty due to photon shot noise for L2F, LDA and DGV is contained in [167], which is based on the Cramér–Rao bound. The comparison was completed by including PTV in [176]. For L2F and LDA, both the Cramér–Rao bound and Heisenberg's uncertainty principle are applied to compare the minimal achievable measurement uncertainty [165]. In the following, the Cramér–Rao bounds for DGV, LDA, L2F and PTV are extracted from [79,165,176]. The measurement uncertainty limits $u(v)$ for DGV, LDA, L2F and PTV, which follow from the square root of the Cramér–Rao bound, are summarized in Table 2.

Table 2. Square root of the Cramér–Rao bound due to photon shot noise for Doppler and time-of-flight measurement techniques, which is the minimal achievable measurement uncertainty $u(v)$ for a single particle and $u(\bar{v})$ for multiple particles, respectively. DGV, Doppler global velocimetry; LDA, laser Doppler anemometer; L2F, laser-2-focus anemometry; PTV, particle tracking velocimetry.

	$u(v)$ for a Single Particle	$u(\bar{v})$ for Multiple Particles	
DGV	$c_1 \cdot \dfrac{\vert v \vert^{1/2} \cdot \frac{\lambda}{\vert \tau' \vert}}{\sqrt{\dot{N}_{\text{photon}}}\, w}$	$c_1 \cdot \dfrac{1}{\sqrt{M}} \cdot \dfrac{\frac{\lambda}{\vert \tau' \vert}}{\sqrt{\dot{N}_{\text{photon}} T}}$	$c_1 = \dfrac{\sqrt{\tau + \tau^2}}{\vert \vec{\sigma} - \vec{i} \vert / \sqrt{2}}$
LDA	$c_2 \cdot \dfrac{\vert v \vert^{3/2}}{\sqrt{\dot{N}_{\text{photon}}}\, w}$	$c_2 \cdot \dfrac{\vert v \vert}{\sqrt{\dot{N}_{\text{photon}} T}}$	$c_2 = \sqrt{\dfrac{3}{\pi}} \cdot \dfrac{d}{w}$
L2F	$c_3 \cdot \dfrac{\vert v \vert^{3/2}}{\sqrt{\dot{N}_{\text{photon}}}\, w}$	$c_3 \cdot \dfrac{\vert v \vert}{\sqrt{\dot{N}_{\text{photon}} T}}$	$c_3 = \dfrac{\tilde{b}}{w}$
PTV	$c_4 \cdot \dfrac{\vert v \vert^{3/2}}{\sqrt{\dot{N}_{\text{photon}}}\, w}$	$c_4 \cdot \dfrac{\vert v \vert}{\sqrt{\dot{N}_{\text{photon}} T}}$	$c_4 = \dfrac{0.94}{w_{\text{px}}}$

The results in the first column in Table 2 are valid for the case of evaluating the scattered light signal from a single scattering particle. In this case, the temporal resolution is determined by the particle transit time $T_t = w/\vert v \vert$ with the particle velocity component v along the sensitivity direction and the respective spatial resolution w, which is the dimension of the measurement volume. As a result, all measurement techniques are indirectly proportional to the square root of the spatial resolution w and the square root of the rate $\dot{N}_{\text{photon}}$ of the observed, scattered photons. Note that the photon rate is defined as an average photon rate by $\dot{N}_{\text{photon}} = N_{\text{photon}}/T_t$, where N_{photon} is the total number of observed, scattered photons per particle. Note also that the product $\dot{N}_{\text{photon}} w$ represents the integral of the available photon rate over space and is proportional to the available light power. Furthermore, the uncertainty limit is directly proportional to the particle velocity v to the power of $\frac{3}{2}$ for LDA, L2F and PTV and to the power of $\frac{1}{2}$ for DGV. Concerning DGV, the ratio $\frac{\lambda}{|\tau'|}$ occurs instead of the factor v. Assuming the typical values for the laser wavelength $\lambda = 532\,\text{nm}$ and the maximal slope $|\tau'| = 2\,\text{GHz}^{-1}$ of the spectral transmission curve of the molecular filter [79], the ratio $\frac{\lambda}{|\tau'|}$ is larger for velocities $v < 266\,\text{m/s}$. Hence, DGV has a lower sensitivity for sub-sonic flows.

In the uncertainty limit, the factors c_1, c_2, c_3, c_4 occur. For DGV, the factor c_1 results from the transmission τ of the molecular filter and the length of the sensitivity vector $\vec{o} - \vec{i}$. With $\tau = 0.5$ and a perpendicular arrangement of the observation and illumination direction ($|\vec{o} - \vec{i}| = \sqrt{2}$), it is $c_1 \approx 0.9$. For LDA, the factor c_2 is mainly determined by the ratio $\frac{d}{w}$ between the distance d of the interference fringes and the dimension w of the measurement volume perpendicular to the fringes ($1/e^2$ width). This ratio is the reciprocal of the number of fringes. For 10 fringes, it is for instance $c_2 \approx 0.1$. For L2F, the factor c_3 is the ratio $\frac{\tilde{b}}{w}$ between the diameter $\tilde{b}$ ($1/e^2$ width) of the two beams and the distance w of the parallel beams. Here, it is assumed $c_3 \approx 0.1$ as an example. For PTV, the factor c_4 is the reciprocal of the particle displacement w_{px} between the two light pulse in unit pixels and can also be understood as a measure of the ratio between an optimal size of the particle image and the particle displacement. For the typical setting $w_{\text{px}} \approx 10$, it is $c_4 \approx 0.1$. As a result, the uncertainty limits for LDA, L2F and PTV are equal and usually lower than for DGV. Assuming a spatial resolution of $w = 100\,\mu\text{m}$ and an averaged scattered light power of 0.1 nW with the wavelength $\lambda = 532\,\text{nm}$, i.e., an average photon rate $\dot{N}_{\text{photon}} = 2.7 \times 10^8\,\text{s}^{-1}$, the measurement uncertainty limit $u(v)$ over the velocity is shown in Figure 9a for all four measurement techniques.

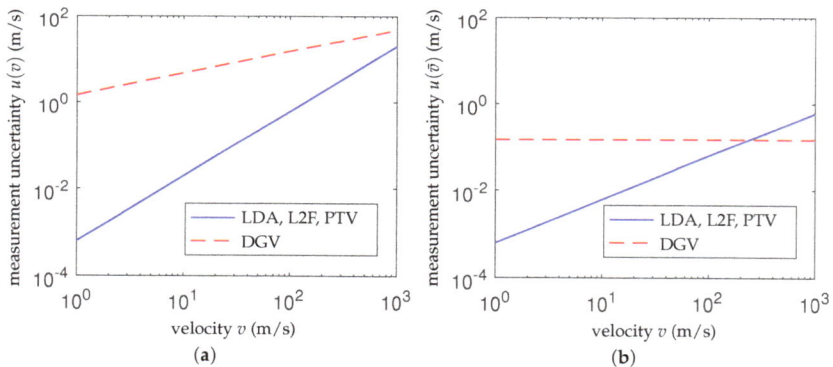

Figure 9. Minimal achievable measurement uncertainty due to photon shot noise over the flow velocity v for an average photon rate per particle of $\dot{N}_{\text{photon}} = 2.7 \times 10^8\,\text{s}^{-1}$ (**a**) for a single particle (that passes the measurement volume with the respective dimension $w = 100\,\mu\text{m}$) and (**b**) for multiple particles (that occur during the measurement time $T = 0.1\,\text{s}$). The formulas are from Table 2 and $c_1 = 0.9$, $c_2 = c_3 = c_4 = 0.1$, $M = 1$, $\lambda = 532\,\text{nm}$, $\tau' = 2\,\text{GHz}^{-1}$.

Note that the same average photon rate and the same spatial resolution are assumed for the different measurement techniques. The same spatial resolution in all dimensions means for L2F that two light sheets instead of two light beams are considered. When using light beams, the average photon rate, as well as the spatial resolution perpendicular to the sensitivity direction is usually higher for L2F than for LDA, DGV and PTV.

The flow velocity measurement can be performed for multiple scattering particles, which pass the measurement volume during a given measurement time T (temporal resolution). The measurement result is then the mean value of the particle velocity. Considering a measurement time T that is significantly larger than the transit time T_t of a single particle through the measurement volume, the maximal number of measured particles is $T/T_t = T/(w/v)$. Hence, the measurement uncertainty is reduced by the factor $\frac{1}{\sqrt{T/(w/v)}}$. The resulting uncertainty limit $u(\bar{v})$ is given in the second column in Table 2. In addition, DGV is not restricted to having one single particle in the measurement volume. The higher the number M of scattering particles in the measurement volume, the larger is the total number of photons that are scattered and observed. Indeed, DGV is known to require a higher seeding particle concentration than other techniques. In order to take this effect into account, M is included in the respective equation and $M = 100$ is considered in the numerical example. The resulting measurement uncertainty limit for DGV is independent of the particle velocity, whereas the measurement uncertainty limit for LDA, L2F and PTV is directly proportional to the particle velocity, cf. Figure 9b for a temporal resolution of $T = 0.1$ ms. Below 239 m/s, the minimal achievable measurement uncertainty $u(\bar{v})$ due to photon shot noise for DGV is higher than for LDA, L2F and PTV. For the example of an averaged scattered light power of 0.1 nW, the measurement uncertainty limit is 0.15 m/s for DGV and 0.06% for LDA, L2F and PTV, respectively.

5. Application Examples, Challenges and Perspectives

Since the applications of ILDV as the imaging Doppler technique with the frequency-based signal are limited to slow convection flows and SFV as imaging time-of-flight techniques with time measurements are more focused on surface velocity measurements, the following consideration is focused on applications of DGV as an imaging Doppler technique with amplitude-based signal evaluation and PIV/PTV as imaging time-of-flight techniques with space measurements. An overview of the manifold PIV applications is given in the book [179]. A textbook that summarizes the manifold DGV applications does not exist, because the absolute number of groups working with DGV is very low. For a comparison between PIV and DGV, we refer the reader to the work of Willert et al. [136], who applied both techniques in large-scale wind tunnels.

Aerodynamic flow studies, e.g., in turbomachinery, as well as fundamental studies of boundary layers and turbulent flows usually require measurements close to walls. Boundary layer measurements close to walls with DGV and PTV are reported in [139,180,181], respectively. Under optimal conditions, PTV measurements allow flow velocity measurements with a minimal distance that is the radius of the scattering particle. DGV and PIV measurements inside a rotating machine are described in [182–184]. The occurrence of unavoidable light reflections causes measurement errors for PIV and DGV [185]. Error correction strategies based on a temporal filter to extract the moving particles [186] and based on the measurement with multiple observation directions to eliminate the influence of the incident direction [139] were developed for PIV and DGV, respectively.

The investigation of flows in combustors and flames is an important application area. DGV measurements in flames were pioneered by Schodl et al. [187]. Combined DGV-PIV measurements in a combustor to obtain 3d velocity images with a single observation direction were introduced by Willert et al. [36]. The single observation direction perpendicular to the window of the combustion chamber minimizes the errors due to image distortions. Furthermore, Fischer et al. reported time-resolved DGV measurements with a measurement rate of 100 kHz to investigate flow velocity oscillations and thermoacoustic phenomena [188,189]. High-speed PIV measurements with rates up to 10 kHz were achieved by Boxx et al. [119,190], whereas volumetric, single shot PIV measurements in flames were for instance reported by Tokarev et al. [191]. Flow measurements in flames are challenging due to the flame light, which disturbs the optical flow velocity measurement. State-of-the-art techniques to cope with this influence are blocking the flame light in the detection unit with bandpass filters while illuminating with narrow-band lasers, applying high power lasers and/or pulsed lasers to surpass the flame light intensity and coding the scattered light signal for instance as a period signal with a high frequency. Furthermore, spatiotemporal variations of the refractive index field occur, which causes measurement errors for PIV and DGV. The effect is qualitatively described for PIV regarding flame flow measurements by Stella et al. [192] and also quantitatively characterized regarding the shock wave in a shear layer and the different air density at an expansion fan by Elsinga et al. [193].

Schlüßler et al. [194] performed simulations and experiments to compare the performance of PIV and DGV for refractive index changes in the imaging path by droplets condensed on a glass window and a flame. DGV seems to be more robust with respect to such disturbances. However, further studies are required to quantify the measurement error for PIV and DGV in flame flows. The challenge to measure in flows with a varying refractive index field also occurs for two-phase flows such as sprays.

Spray flows are important, e.g., to optimize the injection of fuel. PIV spray measurements are described in [195,196]. Zhang et al. reported high-speed PIV measurements with 6 kHz using pulsed illumination [197]. Time-resolved DGV spray measurements were performed with continuous-wave illumination by Fischer et al. [83]. Similar DGV measurements of a high-pressure injection with rates up to 250 kHz were demonstrated by Schlüßler et al. [84] and Gürtler et al. [86]. Due to the continuous illumination and the high speed of the injection, the scattered light intensity, as well as the flow velocity varies during the exposure time of 0.45 µs. In order to improve the temporal resolution by shortening the effective exposure time, high-speed DGV measurements with pulsed lasers are desired. However, short pulses widen the spectrum of the light, which finally deteriorates the sensitivity of DGV if the spectrum is wider than the spectral width of the molecular filter. According to the Fourier limit of laser pulses [198], the typically maximal acceptable bandwidth (full width at half maximum intensity) of $\Delta f = 100$ MHz corresponds to a Gaussian pulse with the duration (full width at half maximum intensity) of $\Delta t = \frac{0.441}{\Delta f} = 4.4$ ns. Hence, the realization, characterization and application of high-speed DGV measurements with pulsed lasers seem feasible and remain to be investigated in future research.

6. Conclusions

Imaging flow velocity measurements with high measurement rates are challenging, because the available light energy needs to be distributed over space and time. Since Mie scattering on seeding/scattering particles provides a higher signal to noise ratio than Rayleigh scattering on the fluid molecules, flow velocity measurement principles based on Mie scattering are advantageous. The respective measurement principles can be subdivided into Doppler and time-of-flight principles. The measurement principles based on the Doppler effect can be further subdivided into principles with amplitude-based and frequency-based signal evaluation. Realizations of these principles with imaging capability are DGV and ILDV, respectively. The time-of-flight principles can be subdivided into principles with time and space measurements. Well-known examples with imaging capability are SFV and PIV/PTV, respectively. All measurement approaches are in principle suitable for imaging flow velocity measurements. However, the Doppler principles with frequency-based signal evaluation and the time-of-flight principles with space measurements require further work to be applicable for the measurement of meso-scale flows. The most advanced field measurement techniques especially regarding a high measurement rate are PIV/PTV and DGV, because a minimum of one or two camera frames are sufficient for one flow velocity measurement. Continuous measurements with up to 100 kHz have been demonstrated, which is sufficient for a wide area of flow applications. Higher measurement rates up to 1 MHz are possible, as well, but are restricted in the number of successive measurements.

Fundamental measurement limits exist due to the measurement at seeding particles, which are the retroaction to the fluid flow, the flow sampling phenomenon, the particle slip, the Brownian particle motion, as well as the illumination-induced force on the particles. An important measurement limit for imaging flow velocity measurements with a high measurement rate occurs due to the quantum noise, i.e., the photon shot noise regarding the number of photons. The measurement uncertainty is indirectly proportional to the square root of the number of photons, which decreases when the available light energy needs to be distributed over space and time. A comparison of the minimal achievable measurement uncertainty shows that Doppler techniques with amplitude-based signal evaluation provide a lower precision than the other Mie-based flow measurement techniques for sub-sonic air flows. However, the image processing requirements are minimal, which is beneficial for real-time imaging flow velocity measurements.

DGV and PIV/PTV are well-developed field measurement tools that have been validated, e.g., for aerodynamic flow studies, near-wall measurements, as well as for two-phase spray flows and flame flows. However, coping with optical flow measurements through scattering media and in fluids with a spatiotemporal variation of the refractive index field, as well as combined measurements of the flow velocity, the temperature and the pressure are topical challenges for Mie-based flow measurements. Furthermore, an accelerated image (pre-)processing either directly on the camera chip or externally with separate processing units allows one to further increase the data rate of the flow velocity imaging techniques in future.

Conflicts of Interest: The authors declare no conflict of interest.

References

1. Fansler, T.D.; Parrish, S.E. Spray measurement technology: A review. *Meas. Sci. Technol.* **2015**, *26*, 012002.
2. Nathan, G.J.; Kalt, P.A.M.; Alwahabi, Z.T.; Dally, B.B.; Medwell, P.R.; Chan, Q.N. Recent advances in the measurement of strongly radiating, turbulent reacting flows. *Progress Energy Combust. Sci.* **2012**, *38*, 41–61.
3. Candel, S.; Durox, D.; Schuller, T.; Bourgouin, J.F.; Moeck, J.P. Dynamics of Swirling Flames. *Ann. Rev. Fluid Mech.* **2014**, *46*, 147–173.
4. Anderson, R.; Zhang, L.; Ding, Y.; Blanco., M.; Bi, X.P.; Wilkinson, D. A critical review of two-phase flow in gas flow channels of proton exchange membrane fuel cells. *J. Power Sources* **2010**, *195*, 4531–4553.
5. Barlas, T.K.; van Kuik, G.A.M. Review of state of the art in smart rotor control research for wind turbines. *Prog. Aerosp. Sci.* **2010**, *46*, 1–27.
6. Tucker, P.G. Computation of unsteady turbomachinery flows: Part 1—Progress and challenges. *Prog. Aerosp. Sci.* **2011**, *47*, 522–545.
7. Pollard, A.; Uddin, M.; Shinneeb, A.M.; Ball, C. Recent advances and key challenges in investigations of the flow inside human oro-pharyngeal-laryngeal airway. *Int. J. Comput. Fluid Dyn.* **2012**, *26*, 363–381.
8. Ford, M.D.; Nikolov, H.N.; Milner, J.S.; Lownie, S.P.; DeMont, E.M.; Kalata, W.; Loth, F.; Holdsworth, D.W.; Steinman, D.A. PIV-Measured Versus CFD-Predicted Flow Dynamics in Anatomically Realistic Cerebral Aneurysm Models. *J. Biomech. Eng.* **2008**, *130*, 021015.
9. Roy, C.J.; Blottner, F.G. Review and assessment of turbulence models for hypersonic flows. *Prog. Aerosp. Sci.* **2006**, *42*, 469–530.
10. Thurow, B.; Jiang, N.; Lempert, W. Review of ultra-high repetition rate laser diagnostics for fluid dynamic measurements. *Meas. Sci. Technol.* **2013**, *24*, 012002.
11. Forkey, J.N.; Finkelstein, N.D.; Lempert, W.R.; Miles, R.B. Demonstration and characterization of filtered Rayleigh scattering for planar velocity measurements. *AIAA J.* **1996**, *34*, 442–448.
12. Seasholtz, R.G.; Buggele, A.E.; Reeder, M.F. Flow Measurements Based on Rayleigh Scattering and Fabry-Perot Interferometer. *Opt. Laser Eng.* **1997**, *27*, 543–570.
13. Grinstead, J.H.; Finkelstein, N.D.; Lempert, W.R. Doppler velocimetry in a supersonic jet by use of frequency-modulated filtered light scattering. *Opt. Lett.* **1997**, *22*, 331–333.
14. Forkey, J.N.; Lempert, W.R.; Miles, R.B. Accuracy limits for planar measurements of flow field velocity, temperature and pressure using Filtered Rayleigh Scattering. *Exp. Fluids* **1998**, *24*, 151–162.
15. Mach, J.; Varghese, P.L. Velocity Measurements by Modulated Filtered Rayleigh Scattering Using Diode Lasers. *AIAA J.* **1999**, *37*, 695–699.
16. Elliott, G.S.; Glumac, N.; Carter, C.D. Molecular filtered Rayleigh scattering applied to combustion. *Meas. Sci. Technol.* **2001**, *12*, 452.
17. Schroll, M.; Doll, U.; Stockhausen, G.; Meier, U.; Willert, C.; Hassa, C.; Bagchi, I. Flow Field Characterization at the Outlet of a Lean Burn Single-Sector Combustor by Laser-Optical Methods. *J. Eng. Gas Turbines Power* **2016**, *139*, 011503.
18. Miles, R.B.; Lempert, W.R.; Forkey, J.N. Laser Rayleigh scattering. *Meas. Sci. Technol.* **2001**, *12*, R33–R51.
19. Doll, U.; Stockhausen, G.; Willert, C. Endoscopic filtered Rayleigh scattering for the analysis of ducted gas flows. *Exp. Fluids* **2014**, *55*, 1690.
20. Doll, U.; Burow, E.; Stockhausen, G.; Willert, C. Methods to improve pressure, temperature and velocity accuracies of filtered Rayleigh scattering measurements in gaseous flows. *Meas. Sci. Technol.* **2016**, *27*, 125204.

21. Mielke, A.F.; Elam, K.A.; Sung, S.J. Multiproperty Measurements at High Sampling Rates Using Rayleigh Scattering. *AIAA J.* **2009**, *47*, 850–862.

22. Chen, L.; Yang, F.R.; Su, T.; Bao, W.Y.; Yan, B.; Chen, S.; Li, R.B. High sampling-rate measurement of turbulence velocity fluctuations in Mach 1.8 Laval jet using interferometric Rayleigh scattering. *Chin. Phys. B* **2017**, *26*, 025205.

23. Mie, G. Beiträge zur Optik trüber Medien, speziell kolloidaler Metalllösungen. *Ann. Phys.* **1908**, *25*, 377–445.

24. Van de Hulst, H.C. *Light Scattering by Small Particles*; Dover Publications: New York, NY, USA, 1981.

25. Bohren, C.F.; Huffman, D.R. *Absorption and Scattering of Light by Small Particles*; Wiley-VCH Verlag: Weinheim, Germany, 2004.

26. Albrecht, H.E.; Damaschke, N.; Borys, M.; Tropea, C. *Laser Doppler and Phase Doppler Measurement Techniques*; Springer: Berlin, Germany, 2003.

27. Tu, C.; Yin, Z.; Lin, J.; Bao, F. A Review of Experimental Techniques for Measuring Micro- to Nano-Particle-Laden Gas Flows. *Appl. Sci.* **2017**, *7*, 120.

28. Charrett, T.O.H.; James, S.W.; Tatam, R.P. Optical fibre laser velocimetry: A review. *Meas. Sci. Technol.* **2012**, *23*, 032001.

29. Einstein, A. Zur Elektrodynamik bewegter Körper. *Ann. Phys. Chem.* **1905**, *17*, 891–921.

30. Drain, L.E. *The Laser Doppler Technique*; John Wiley & Sons: Chichester, UK, 1980.

31. Adrian, R.J. Particle-Imaging Techniques for Experimental Fluid Mechanics. *Ann. Rev. Fluid Mech.* **1991**, *23*, 261–304.

32. Hain, R.; Kähler, C.J. Fundamentals of multiframe particle image velocimetry (PIV). *Exp. Fluids* **2007**, *42*, 575–587.

33. Cierpka, C.; Lütke, B.; Kähler, C.J. Higher order multi-frame particle tracking velocimetry. *Exp. Fluids* **2013**, *54*, 1533.

34. Förster, W.; Karpinsky, G.; Krain, H.; Röhle, I.; Schodl, R. 3-Component-Doppler-Laser-Two-Focus velocimetry applied to a transonic centrifugal compressor. In Proceedings of the 10th International Symposium on Applications of Laser Techniques to Fluid Mechanics; Lisbon, Portugal, 10–13 July 2000; Number 7.2, pp. 1–12.

35. Wernet, M.P. Planar particle imaging Doppler velocimetry: A hybrid PIV/DGV technique for three-component velocity measurements. *Meas. Sci. Technol.* **2004**, *15*, 2011–2028.

36. Willert, C.; Hassa, C.; Stockhausen, G.; Jarius, M.; Voges, M.; Klinner, J. Combined PIV and DGV applied to a pressurized gas turbine combustion facility. *Meas. Sci. Technol.* **2006**, *17*, 1670–1679.

37. Grosjean, N.; Graftieaux, L.; Michard, M.; Hubner, W.; Tropea, C.; Volkert, J. Combining LDA and PIV for turbulence measurements in unsteady swirling flows. *Meas. Sci. Technol.* **1997**, *8*, 1523–1532.

38. Lira, I. *Evaluating the Measurement Uncertainty: Fundamentals and Practical Guidance*; Institute of Physics Publishing: Bristol, PA, USA, 2002.

39. Joint Committee for Guides in Metrology (JCGM). Evaluation of Measurement Data—Guide to the Expression of Uncertainty in Measurement, 2008. Available online: www.bipm.org/en/publications/guides/gum.html (accessed on 9 November 2017).

40. Jackson, D.A.; Paul, D.M. Measurement of hypersonic velocities and turbulence by direct spectral analysis of Doppler shifted laser light. *Phys. Lett.* **1970**, *32A*, 77–78.

41. Kentischer, T.J.; Schmidt, W.; Sigwarth, M.; Uexkull, M.V. TESOS, a double Fabry-Perot instrument for solar spectroscopy. *Astron. Astrophys.* **1998**, *340*, 569–578.

42. Büttner, L.; Schlüßler, R.; Fischer, A.; Czarske, J. Multipoint velocity measurements in flows using a Fabry-Pérot interferometer. *Opt. Lasers Eng.* **2013**, *51*, 325–333.

43. Chehura, E.; Ye, C.C.; Tatam, R.P. In-line laser Doppler velocimeter using fibre-optic Bragg grating interferometric filters. *Meas. Sci. Technol.* **2003**, *14*, 724–735.

44. Smeets, G.; George, A. Instantaneous laser Doppler velocimeter using a fast wavelength tracking Michelson interferometer. *Rev. Sci. Instrum.* **1978**, *49*, 1589.

45. Smeets, G.; George, A. Michelson spectrometer for instantaneous Doppler velocity measurements. *J. Phys. E Sci. Instrum.* **1981**, *14*, 838–845.

46. Oertel, H.; Seiler, F.; George, A. *Visualisierung von Geschwindigkeitsfeldern mit Dopplerbildern*; Visualization of Velocity Fields With Doppler Pictures; ISL-report R 115/82; Springer: Berlin, Germany, 1982.

47. Seiler, F.; Oertel, H. Visualization of velocity fields with Doppler-pictures. In Proceedings of the 3rd International Symposium on Flow Visualization, Ann Arbor, MI, USA , 6–9 September 1983.

48. Seiler, F.; George, A.; Srulijes, J.; Havermann, M. Progress in Doppler picture velocimetry (DPV). *Exp. Fluids* **2008**, *44*, 389–395.

49. Landolt, A.; Roesgen, T. Anomalous dispersion in atomic line filters applied for spatial frequency detection. *Appl. Opt.* **2009**, *48*, 5948–5955.

50. Landolt, A.; Rösgen, T. Global Doppler frequency shift detection with near-resonant interferometry. *Exp. Fluids* **2009**, *47*, 733–743.

51. Lu, Z.H.; Charett, T.O.H.; Ford, H.D.; Tatam, R.P. Mach-Zehnder interferometric filter based planar Doppler velocimetry (MZI-PDV). *J. Opt. A Pure Appl. Opt.* **2007**, *9*, 1002–1013.

52. Lu, Z.H.; Charett, T.O.H.; Tatam, R.P. Three-component planar velocity measurements using Mach-Zehnder interferometric filter-based planar Doppler velocimetry (MZI-PDV). *Meas. Sci. Technol.* **2009**, *20*, 034019.

53. Komine, H. System for Measuring Velocity Field of Fluid Flow Utilizing a Laser-Doppler Spectral Image Converter. U.S. Patent 4,919,536, 24 April 1990.

54. Meyers, J.F. Development of Doppler global velocimetry as a flow diagnostic tool. *Meas. Sci. Technol.* **1995**, *6*, 769–783.

55. Ainsworth, R.W.; Thorpe, S.J.; Manners, R.J. A new approach to flow-field measurement—A view of Doppler global velocimetry techniques. *Int. J. Heat Fluid Flow* **1997**, *18*, 116–130.

56. Nobes, D.S.; Ford, H.D.; Tatam, R.P. Instantaneous, three-component planar Doppler velocimetry using imaging fibre bundles. *Exp. Fluids* **2004**, *36*, 3–10.

57. Röhle, I.; Willert, C.E. Extension of Doppler global velocimetry to periodic flows. *Meas. Sci. Technol.* **2001**, *12*, 420–431.

58. Charrett, T.O.H.; Bledowski, I.A.; James, S.W.; Tatam, R.P. Frequency division multiplexing for interferometric planar Doppler velocimetry. *Appl. Opt.* **2014**, *53*, 4363–4374.

59. Willert, C.; Stockhausen, G.; Klinner, J.; Lempereur, C.; Barricau, P.; Loiret, P.; Raynal, J.C. Performance and accuracy investigations of two Doppler global velocimetry systems applied in parallel. *Meas. Sci. Technol.* **2007**, *18*, 2504–2512.

60. Thurow, B.; Jiang, N.; Lempert, W.; Samimy, M. MHz rate planar Doppler velocimetry in supersonic jets. In Proceedings of the 42nd AIAA Aerospace Sciences Meeting and Exhibit, Reno, NV, USA, 6–9 January 2004; Number AIAA-2004-0023.

61. Bloom, S.H.; Kremer, R.; Searcy, P.A.; Rivers, M.; Menders, J.; Korevaar, E. Long-range, noncoherent laser Doppler velocimeter. *Opt. Lett.* **1991**, *16*, 1794–1796.

62. Crafton, J.; Messersmith, N.M.; Sullivan, J.P. Filtered Doppler Velocimetry: Development of a Point System. In Proceedings of the 36th Aerospace Sciences Meeting & Exhibit, Reno, NV, USA, 12–15 January 1998; Number AIAA-98-0509.

63. Arnette, S.A.; Samimy, M.; Elliot, G.S. Two-component planar Doppler velocimetry in the compressible turbulent boundary layer. *Exp. Fluids* **1998**, *24*, 323–332.

64. Charrett, T.O.H.; Ford, H.D.; Nobes, D.S.; Tatam, R.P. Two-Frequency Planar Doppler Velocimetry (2-*v*-PDV). *Rev. Sci. Instrum.* **2004**, *75*, 4487–4496.

65. Charrett, T.O.H.; Tatam, R.P. Single camera three component planar velocity measurements using two-frequency planar Doppler velocimetry (2*v*-PDV). *Meas. Sci. Technol.* **2006**, *17*, 1194–1206.

66. Müller, H.; Eggert, M.; Pape, N.; Dopheide, D.; Czarske, J.; Büttner, L.; Razik, T. Time resolved DGV based on laser frequency modulation. In Proceeding of the 12th International Symposium on Applications of Laser Techniques to Fluid Mechanics, Lisbon, Portugal, 12–15 July 2004; Number 25.2, 10p.

67. Müller, H.; Eggert, M.; Czarske, J.; Büttner, L.; Fischer, A. Single-camera Doppler global velocimetry based on frequency modulation techniques. *Exp. Fluids* **2007**, *43*, 223–232.

68. Eggert, M.; Müller, H.; Czarske, J.; Büttner, L.; Fischer, A. Self-calibrating Single Camera Doppler Global Velocimetry based on Frequency Shift Keying. In *Imaging Measurement Methods for Flow Analysis*; Nitsche, W., Dobriloff, C., Eds.; Springer: Berlin, Germany, 2009; pp. 43–52.

69. Eggert, M.; Müller, H.; Czarske, J.; Büttner, L.; Fischer, A. Self calibrating FSK-Doppler global velocimetry for three-componential time resolved and phase averaged flow field measurements. In Proceedings of the 15th International Symposium on Applications of Laser Techniques to Fluid Mechanics, Lisbon, Portugal, 7–10 July 2010; Number 1.11.1, 11p.

70. Müller, H.; Lehmacher, H.; Grosche, G. Profile sensor based on Doppler Global Velocimetry. In Proceedings of the 8th International Conference Laser Anemometry Advances and Applications, Rome, Italy, 6–8 September 1999; pp. 475–482.

71. Müller, H.; Pape, N.; Grosche, G.; Strunck, V.; Dopheide, D. Simplified DGV on-line profile sensor. In Proceedings of the 11th International Symposium on Applications of Laser Techniques to Fluid Mechanics, Lisbon, Portugal, 8–11 July 2002; Number 9.3, 8p.

72. Fischer, A.; Büttner, L.; Czarske, J.; Eggert, M.; Grosche, G.; Müller, H. Investigation of time-resolved single detector Doppler global velocimetry using sinusoidal laser frequency modulation. *Meas. Sci. Technol.* **2007**, *18*, 2529–2545.

73. Fischer, A.; Büttner, L.; Czarske, J.; Eggert, M.; Müller, H. Measurement uncertainty and temporal resolution of Doppler global velocimetry using laser frequency modulation. *Appl. Opt.* **2008**, *47*, 3941–3953.

74. Fischer, A.; Büttner, L.; Czarske, J.; Eggert, M.; Müller, H. Array Doppler global velocimeter with laser frequency modulation for turbulent flow analysis—Sensor investigation and application. In *Imaging Measurement Methods for Flow Analysis*; Nitsche, W., Dobriloff, C., Eds.; Springer: Berlin, Germany, 2009; pp. 31–41.

75. Cadel, D.R.; Lowe, K.T. Cross-correlation Doppler global velocimetry (CC-DGV). *Opt. Lasers Eng.* **2015**, *71*, 51–61.

76. Cadel, D.R.; Lowe, K.T. Investigation of measurement sensitivities in cross-correlation Doppler global velocimetry. *Opt. Lasers Eng.* **2016**, *86*, 44–52.

77. Fischer, A.; Büttner, L.; Czarske, J. Simultaneous measurements of multiple flow velocity components using frequency modulated lasers and a single molecular absorption cell. *Opt. Commun.* **2011**, *284*, 3060–3064.

78. Fischer, A.; Kupsch, C.; Gürtler, J.; Czarske, J. High-speed light field camera and frequency division multiplexing for fast multi-plane velocity measurements. *Opt. Express* **2015**, *23*, 24910.

79. Fischer, A. Model-based review of Doppler global velocimetry techniques with laser frequency modulation. *Opt. Lasers* **2017**, *93*, 19–35.

80. Fischer, A.; Büttner, L.; Czarske, J.; Eggert, M.; Müller, H. Measurements of velocity spectra using time-resolving Doppler global velocimetry with laser frequency modulation and a detector array. *Exp. Fluids* **2009**, *47*, 599–611.

81. Fischer, A.; König, J.; Haufe, D.; Schlüßler, R.; Büttner, L.; Czarske, J. Optical multi-point measurements of the acoustic particle velocity with frequency modulated Doppler global velocimetry. *J. Acoust. Soc. Am.* **2013**, *134*, 1102–1111.

82. Fischer, A.; Wilke, U.; Schlüßler, R.; Haufe, D.; Sandner, T.; Czarske, J. Extension of frequency modulated Doppler global velocimetry for the investigation of unsteady spray flows. *Opt. Lasers Eng.* **2014**, *63*, 1–10.

83. Fischer, A.; Schlüßler, R.; Haufe, D.; Czarske, J. Lock-in spectroscopy employing a high-speed camera and a micro-scanner for volumetric investigations of unsteady flows. *Opt. Lett.* **2014**, *39*, 5082–5085.

84. Schlüßler, R.; Gürtler, J.; Czarske, J.; Fischer, A. Planar near-nozzle velocity measurements during a single high-pressure fuel injection. *Exp. Fluids* **2015**, *56*, 176.

85. Gürtler, J.; Haufe, D.; Schulz, A.; Bake, F.; Enghardt, L.; Czarske, J.; Fischer, A. High-speed camera-based measurement system for aeroacoustic investigations. *J. Sens. Sens. Syst.* **2016**, *5*, 125–136.

86. Gürtler, J.; Schlüßler, R.; Fischer, A.; Czarske, J. High-speed non-intrusive measurements of fuel velocity fields at high pressure injectors. *Opt. Lasers Eng.* **2017**, *90*, 91–100.

87. Yeh, Y.; Cummins, H.Z. Localized Fluid Flow Measurements with an He-Ne Laser Spectrometer. *Appl. Phys. Lett.* **1964**, *4*, 176–178.

88. Li, C.H.; Benedick, A.J.; Fendel, P.; Glenday, A.G.; Kartner, F.X.; Phillips, D.F.; Sasselov, D.; Szentgyorgyi, A.; Walsworth, R.L. A laser frequency comb that enables radial velocity measurements with a precision of 1 cm s^{-1}. *Nature* **2008**, *452*, 610–612.

89. Coupland, J. Coherent detection in Doppler global velocimetry: a simplified method to measure subsonic fluid flow fields. *Appl. Opt.* **2000**, *39*, 1505–1510.

90. Meier, A.H.; Rösgen, T. Heterodyne Doppler global velocimetry. *Exp. Fluids* **2009**, *47*, 665–672.

91. Tropea, C. Laser Doppler anemometry: recent developments and future challenges. *Meas. Sci. Technol.* **1995**, *6*, 605–619.
92. Czarske, J.W. Laser Doppler velocimetry using powerful solid-state light sources. *Meas. Sci. Technol.* **2006**, *17*, R71–R91.
93. Czarske, J. Laser Doppler velocity profile sensor using a chromatic coding. *Meas. Sci. Technol.* **2001**, *12*, 52–57.
94. Czarske, J.; Büttner, L.; Razik, T.; Müller, H. Boundary layer velocity measurements by a laser Doppler profile sensor with micrometre spatial resolution. *Meas. Sci. Technol.* **2002**, *13*, 1979–1989.
95. Voigt, A.; Bayer, C.; Shirai, K.; Büttner, L.; Czarske, J. Laser Doppler field sensor for high resolution flow velocity imaging without camera. *Appl. Opt.* **2008**, *47*, 5028–5040.
96. Meier, A.H.; Rösgen, T. Imaging laser Doppler velocimetry. *Exp. Fluids* **2012**, *52*, 1017–1026.
97. Thompson, D.H. A tracer-particle fluid velocity meter incorporating a laser. *J. Phys. E Sci. Instrum.* **1968**, *1*, 929–932.
98. Tanner, L. A particle timing laser velocity meter. *Opt. Laser Technol.* **1973**, *5*, 108–110.
99. Schodl, R. A Laser-Two-Focus (L2F) Velocimeter for Automatic Flow Vector Measurements in the Rotating Components of Turbomachines. *J. Fluids Eng.* **1980**, *102*, 412–419.
100. Ator, J.T. Image-velocity sensing with parallel-slit reticles. *J. Opt. Soc. Am.* **1963**, *53*, 1416–1419.
101. Gaster, M. A new technique for the measurement of low fluid velocities. *J. Fluid Mech.* **1964**, *20*, 183–192.
102. Aizu, Y.; Asakura, T. Principles and development of spatial filtering velocimetry. *Appl. Phys. B Lasers Opt.* **1987**, *43*, 209–224.
103. Aizu, Y.; Asakura, T. *Spatial Filtering Velocimetry: Fundamentals and Applications*; Springer: Berlin, Germany, 2006.
104. Christofori, K.; Michel, K. Velocimetry with spatial filters based on sensor arrays. *Flow Meas. Instrum.* **1996**, *7*, 265–272.
105. Michel, K.C.; Fiedler, O.F.; Richter, A.; Christofori, K.; Bergeler, S. A Novel Spatial Filtering Velocimeter Based on a Photodetector Array. *IEEE Trans. Instrum. Meas.* **1998**, *47*, 299–304.
106. Bergeler, S.; Krambeer, H. Novel optical spatial filtering methods based on two-dimensional photodetector arrays. *Meas. Sci. Technol.* **2004**, *15*, 1309–1315.
107. Schaeper, M.; Damaschke, N. Fourier-based layout for grating function structure in spatial filtering velocimetry. *Meas. Sci. Technol.* **2017**, *28*, 055008.
108. Pau, S.; Dallas, W.J. Generalized spatial filtering velocimetry and accelerometry for uniform and nonuniform objects. *Appl. Opt.* **2009**, *48*, 4713–4722.
109. Hosokawa, S.; Mastumoto, T.; Tomiyama, A. Tomographic spatial filter velocimetry for three-dimensional measurement of fluid velocity. *Exp. Fluids* **2013**, *54*, 1597.
110. Schaeper, M.; Damaschke, N. Velocity Measurement for Moving Surfaces by Using Spatial Filtering Technique Based on Array Detectors. In Proceedings of the AIS: International Conference on Autonomous and Intelligent Systems, Povoa de Varzim, Portugal, 21–23 June 2011; pp. 303–310.
111. Adrian, R.J. Scattering particle characteristics and their effect on pulsed laser measurements of fluid flow: Speckle velocimetry vs. particle image velocimetry. *Appl. Opt.* **1984**, *23*, 1690–1691.
112. Adrian, R.J. Twenty years of particle image velocimetry. *Exp. Fluids* **2005**, *39*, 159–169.
113. Raffel, M.; Willert, C.E.; Scarano, F.; Kähler, C.J.; Wereley, S.T.; Kompenhans, J. *Particle Image Velocimetry*; Springer: Berlin, Germany, 2018.
114. Willert, C.; Stasicki, B.; Klinner, J.; Moessner, S. Pulsed operation of high-power light emitting diodes for imaging flow velocimetry. *Meas. Sci. Technol.* **2010**, *21*, 075402.
115. Wernet, M.P. Temporally resolved PIV for space-time correlations in both cold and hot jet flows. *Meas. Sci. Technol.* **2007**, *18*, 1387–1403.
116. Arroyo, M.P.; Greated, C.A. Stereoscopic particle image velocimetry. *Meas. Sci. Technol.* **1991**, *2*, 1181–1186.
117. Boxx, I.; Stöhr, M.; Carter, C.; Meier, W. Sustained multi-kHz flamefront and 3-component velocity-field measurements for the study of turbulent flames. *Appl. Phys. B* **2009**, *95*, 23–29.
118. Boxx, I.; Arndt, C.M.; Carter, C.D.; Meier, W. High-speed laser diagnostics for the study of flame dynamics in a lean premixed gas turbine model combustor. *Exp. Fluids* **2012**, *52*, 555–567.
119. Boxx, I.; Carter, C.D.; Stöhr, M.; Meier, W. Study of the mechanisms for flame stabilization in gas turbine model combustors using kHz laser diagnostics. *Exp. Fluids* **2013**, *54*, 1532.
120. Scarano, F. Tomographic PIV: Principles and practice. *Meas. Sci. Technol.* **2013**, *24*, 012001.

121. Hinsch, K.D. Holographic particle image velocimetry. *Meas. Sci. Technol.* **2002**, *13*, R61–R72.
122. Buchmann, N.A.; Atkinson, C.; Soria, J. Ultra-high-speed tomographic digital holographic velocimetry in supersonic particle-laden jet flows. *Meas. Sci. Technol.* **2013**, *24*, 024005.
123. Cenedese, A.; Cenedese, C.; Furia, F.; Marchetti, M.; Moroni, M.; Shindler, L. 3D particle reconstruction using light field imaging. In Proceedings of the 16th International Symposium on Applications of Laser Techniques to Fluid Mechanics, Lisbon, Portugal, 9–12 July 2012; Number 1.1.2, 9p.
124. Belden, J.; Truscott, T.T.; Axiak, M.C.; Techet, A.H. Three-dimensional synthetic aperture particle image velocimetry. *Meas. Sci. Technol.* **2010**, *21*, 125403.
125. Fahringer, T.W.; Lynch, K.P.; Thurow, B.S. Volumetric particle image velocimetry with a single plenoptic camera. *Meas. Sci. Technol.* **2015**, *26*, 115201.
126. Deem, E.; Zhang, Y.; Cattafesta, L.; Fahringer, T.; Thurow, B. On the resolution of plenoptic PIV. *Meas. Sci. Technol.* **2016**, *27*, 084003.
127. Wernet, M.P. Two-dimensional particle displacement tracking in particle imaging velocimetry. *Appl. Opt.* **1991**, *30*, 1839–1846.
128. Maas, H.G.; Gruen, A.; Papantoniou, D. Particle tracking velocimetry in three-dimensional flows—Part 1: Photogrammetric determination of particle coordinates. *Exp. Fluids* **1993**, *15*, 133–146.
129. Malik, N.A.; Dracos, T.; Papantoniou, D.A. Particle tracking velocimetry in three-dimensional flows—Part II: Particle tracking. *Exp. Fluids* **1993**, *15*, 279–294.
130. Willert, C.E.; Gharib, M. Three-dimensional particle imaging with a single camera. *Exp. Fluids* **1992**, *12*, 353–358.
131. Stolz, W.; Köhler, J. In-plane determination of 3D-velocity vectors using particle tracking anemometry (PTA). *Exp. Fluids* **1994**, *17*, 105–109.
132. Cierpka, C.; Segura, R.; Hain, R.; Köhler, C.J. A simple single camera 3C3D velocity measurement technique without errors due to depth of correlation and spatial averaging for microfluidics. *Meas. Sci. Technol.* **2010**, *21*, 045401.
133. Kreizer, M.; Ratner, D.; Liberzon, A. Real-time image processing for particle tracking velocimetry. *Exp. Fluids* **2010**, *48*, 105–110.
134. Kreizer, M.; Liberzon, A. Three-dimensional particle tracking method using FPGA-based real-time image processing and four-view image splitter. *Exp. Fluids* **2011**, *50*, 613–620.
135. Buchmann, N.A.; Cierpka, C.; Köhler, C.J.; Soria, J. Ultra-high-speed 3D astigmatic particle tracking velocimetry: Application to particle-laden supersonic impinging jets. *Exp. Fluids* **2014**, *55*, 1842.
136. Willert, C.; Stockhausen, G.; Beversdorff, M.; Klinner, J.; Lempereur, C.; Barricau, P.; Quest, J.; Jansen, U. Application of Doppler global velocimetry in cryogenic wind tunnels. *Exp. Fluids* **2005**, *39*, 420–430.
137. White, F.M. *Viscous Fluid Flow*, 2nd ed.; McGraw-Hill: New York, NY, USA, 1991.
138. McKenzie, R.L. Measurement capabilities of planar Doppler velocimetry using pulsed lasers. *Appl. Opt.* **1996**, *35*, 948–964.
139. Fischer, A.; Haufe, D.; Büttner, L.; Czarske, J. Scattering effects at near-wall flow measurements using Doppler global velocimetry. *Appl. Opt.* **2011**, *50*, 4068–4082.
140. Shirai, K.; Pfister, T.; Büttner, L.; Czarske, J.; Müller, H.; Becker, S.; Lienhart, H.; Durst, F. Highly spatially resolved velocity measurements of a turbulent channel flow by a fiber-optic heterodyne laser-Doppler velocity-profile sensor. *Exp. Fluids* **2006**, *40*, 473–481.
141. Woisetschläger, J.; Göttlich, E. Recent Applications of Particle Image Velocimetry to Flow Research in Thermal Turbomachinery. In *Particle Image Velocimetry*; Topics in Applied Physics; Springer: Berlin/ Heidelberg, Germany, 2008; Volume 112, pp. 311–331.
142. Pope, S.B. *Turbulent Flows*; Cambridge University Press: Cambridge, UK, 2000.
143. Durst, F.; Martinuzzi, R.; Sender, J.; Thevenin, D. LDA-Measurements of Mean Velocity, RMS-Values and Higher Order Moments of Turbulence Intensity Fluctuations in Flow Fields with Strong Velocity Gradients. In Proceedings of the 6th International Symposium on Applications of Laser Techniques to Fluid Mechanics, Lisbon, Portugal, 20–23 July 1992; Number S5-1, 6p.
144. Durst, F.; Jovanović, J.; Sender, J. LDA measurements in the near-wall region of a turbulent pipe flow. *J. Fluid Mech.* **1995**, *295*, 305–355.
145. Fischer, M.; Jovanović, J.; Durst, F. Near-wall behaviour of statistical properties in turbulent Fows. *Int. J. Heat Fluid Flow* **2000**, *21*, 471–479.

146. Nobach, H. Influence of individual variations of particle image intensities on high-resolution PIV. *Exp. Fluids* **2011**, *50*, 919–927.

147. Basset, A.B. *Treatise on Hydrodynamics*; Deighton, Bell & Co.: Cambridge, UK, 1888; Volume 2.

148. Saffman, P.G. The lift on a small sphere in a slow shear flow. *J. Fluid Mech.* **1965**, *22*, 385–400.

149. Durst, F.; Melling, A.; Whitelaw, J.H. *Principles and Practice of Laser-Doppler Anemometry*; Academic Press: London, UK, 1981.

150. Mei, R. Velocity fidelity of flow tracer particles. *Exp. Fluids* **1996**, *22*, 1–13.

151. Crowe, C.; Schwarzkopf, J.D.; Sommerfeld, M.; Tsuji, Y. *Multiphase Flows with Droplets and Particles*, 2nd ed.; CRC Press: Boca Raton, FL, USA, 2011.

152. Melling, A. Tracer particles and seeding for particle image velocimetry. *Meas. Sci. Technol.* **1997**, *8*, 1406–1416.

153. Hjelmfelt, A.T.; Mockros, L.F. Motion of discrete particles in a turbulent fluid. *App. Sci. Res.* **1966**, *16*, 149–161.

154. Uhlenbeck, G.E.; Ornstein, L.S. On the theory of Brownian motion. *Phys. Rev.* **1930**, *36*, 823–841.

155. Wang, M.C.; Uhlenbeck, G.E. On the theory of Brownian motion II. *Rev. Mod. Phys.* **1945**, *17*, 323–342.

156. Coffey, W.T.; Kalmykov, Y.P.; Waldron, J.T. *The Langevin Equation: With Applications in Physics, Chemistry, and Electrical Engineering*; World Scientific Publishing: Singapore, 1996.

157. Durst, F.; Ruck, B. Influence of laser radiation on particle properties, Part 1: Influence of radiation pressure on the particle motion. *tm—Technisches Messen* **1980**, *47*, 233–230. (In German)

158. Durst, F.; Ruck, B. Influence of laser radiation on particle properties, Part 2: Particle heating by radiation absorption. *tm—Technisches Messen* **1980**, *47*, 267–272. (In German)

159. Saleh, B.E.A.; Teich, M.C. *Fundamentals of Photonics*; John Wiley & Sons: New York, NY, USA, 2007.

160. Rao, C.R. Information and the accuracy attainable in the estimation of statistical parameters. *Bull. Calcutta Math. Soc.* **1945**, *37*, 81–91.

161. Cramer, H. *Mathematical Methods of Statistics*; Princeton University Press: Princeton, NJ, USA, 1946.

162. Schervish, M.J. *Theory of Statistics*; Springer: Berlin, Germany, 1997.

163. Casella, G.; Berger, R.L. *Statistical Inference*; Duxbury Press: Belmont, CA, USA, 1990.

164. Heisenberg, W. Über den anschaulichen Inhalt der quantentheoretischen Kinematik und Mechanik. *Z. Phys.* **1927**, *43*, 172–198.

165. Fischer, A. Fundamental uncertainty limit of optical flow velocimetry according to Heisenberg's uncertainty principle. *Appl. Opt.* **2016**, *55*, 8787–8795.

166. Fischer, A.; Czarske, J. Signal processing efficiency of Doppler global velocimetry with laser frequency modulation. *Opt. Int. J. Light Electron Opt.* **2010**, *121*, 1891–1899.

167. Fischer, A.; Pfister, T.; Czarske, J. Derivation and comparison of fundamental uncertainty limits for laser-two-focus velocimetry, laser Doppler anemometry and Doppler global velocimetry. *Measurement* **2010**, *43*, 1556–1574.

168. Rife, D.C.; Boorstyn, R.R. Single-tone parameter estimation from discrete-time observations. *IEEE Trans. Inf. Theory* **1974**, *20*, 591–598.

169. Besson, O.; Galtier, F. Estimating Particles Velocity from Laser Measurements: Maximum Likelihood and Cramér-Rao Bounds. *IEEE Trans. Signal Process.* **1996**, *12*, 3056–3068.

170. Shu, W.Q. Cramér-Rao Bound of Laser Doppler Anemometer. *IEEE Trans. Instrum. Meas.* **2001**, *50*, 1770–1772.

171. Sobolev, V.S.; Feshenko, A.A. Accurate Cramer-Rao Bounds for a Laser Doppler Anemometer. *IEEE Trans. Instrum. Meas.* **2006**, *55*, 659–665.

172. Oliver, C.J. Accuracy in laser anemometry. *J. Phys. D Appl. Phys.* **1980**, *13*, 1145–1159.

173. Lading, L. Estimating time and time-lag in time-of-flight velocimetry. *Appl. Opt.* **1983**, *22*, 3637–3643.

174. Lading, L.; Jørgensen, T.M. Maximizing the information transfer in a quantum-limited light-scattering system. *J. Opt. Soc. Am. A* **1990**, *7*, 1324–1331.

175. Wernet, M.P.; Pline, A. Particle displacement tracking technique and Cramer-Rao lower bound error in centroid estimates from CCD imagery. *Exp. Fluids* **1993**, *15*, 295–307.

176. Fischer, A. *Messbarkeitsgrenzen optischer Strömungsmessverfahren: Theorie und Anwendungen*; Shaker: Aachen, Germany, 2013. (In German)

177. Westerweel, J. Fundamentals of digital particle image velocimetry. *Meas. Sci. Technol.* **1997**, *8*, 1379–1392.

178. Westerweel, J. Theoretical analysis of the measurement precision in particle image velocimetry. *Exp. Fluids* **2000**, *29* (Suppl. S1), S3–S12.

179. Schröder, A.; Willert, C.E. *Particle Image Velocimetry: New Developments and Recent Applications*; Springer: Berlin, Germany, 2008.
180. Meyers, J.F.; Lee, J.W.; Cavone, A.A. Boundary layer measurements in a supersonic wind tunnel using Doppler global velocimetry. In Proceedings of the 15th International Symposium on Applications of Laser Techniques to Fluid Mechanics, Lisbon, Portugal, 7–10 July 2010; Number 1.8.1, 11p.
181. Kähler, C.J.; Scharnowski, S.; Cierpka, C. On the uncertainty of digital PIV and PTV near walls. *Exp. Fluids* **2012**, *52*, 1641–1656.
182. Fischer, A.; König, J.; Czarske, J.; Rakenius, C.; Schmid, G.; Schiffer, H.P. Investigation of the tip leakage flow at turbine rotor blades with squealer. *Exp. Fluids* **2013**, *54*, 1462.
183. Voges, M.; Willert, C.; Mönig, R.; Müller, M.W.; Schiffer, H.P. The challenge of stereo PIV measurements in the tip gap of a transonic compressor rotor with casing treatment. *Exp. Fluids* **2010**, *52*, 581–590.
184. Voges, M.; Schnell, R.; Willert, C.; Mönig, R.; Müller, M.W.; Zscherp, C. Investigation of Blade Tip Interaction With Casing Treatment in a Transonic Compressor – Part I: Particle Image Velocimetry. *J. Turbomach.* **2011**, *133*, 011007.
185. Schlüßler, R.; Blechschmidt, C.; Czarske, J.; Fischer, A. Optimizations for optical velocity measurements in narrow gaps. *Opt. Eng.* **2013**, *52*, 094101.
186. Sciacchitano, A.; Scarano, F. Elimination of PIV light reflections via a temporal high pass filter. *Meas. Sci. Technol.* **2014**, *25*, 084009.
187. Schodl, R.; Röhle, I.; Willert, C.; Fischer, M.; Heinze, J.; Laible, C.; Schilling, T. Doppler global velocimetry for the analysis of combustor flows. *Aerosp. Sci. Technol.* **2002**, *6*, 481–493.
188. Fischer, A.; König, J.; Czarske, J.; Peterleithner, J.; Woisetschläger, J.; Leitgeb, T. Analysis of flow and density oscillations in a swirl-stabilized flame employing highly resolving optical measurement techniques. *Exp. Fluids* **2013**, *54*, 1622.
189. Schlüßler, R.; Bermuske, M.; Czarske, J.; Fischer, A. Simultaneous three-component velocity measurements in a swirl-stabilized flame. *Exp. Fluids* **2015**, *56*, 183.
190. Stöhr, M.; Boxx, I.; Carter, C.; Meier, W. Experimental study of vortex-flame interaction in a gas turbine model combustor. *Combust. Flame* **2012**, *159*, 2636–2649.
191. Tokarev, M.P.; Sharaborin, D.K.; Lobasov, A.S.; Chikishev, L.M.; Dulin, V.M.; Markovich, D.M. 3D velocity measurements in a premixed flame by tomographic PIV. *Meas. Sci. Technol.* **2015**, *26*, 064001.
192. Stella, A.; Guj, G.; Kompenhans, J.; Raffel, M.; Richard, H. Application of particle image velocimetry to combusting flows: Design considerations and uncertainty assessment. *Exp. Fluids* **2001**, *30*, 167–180.
193. Elsinga, G.E.; van Oudheusden, B.W.; Scarano, F. Evaluation of aero-optical distortion effects in PIV. *Exp. Fluids* **2005**, *39*, 246–256.
194. Schlüßler, R.; Czarske, J.; Fischer, A. Uncertainty of flow velocity measurements due to refractive index fluctuations. *Opt. Lasers Eng.* **2014**, *54*, 93–104.
195. Cao, Z.M.; Nishino, K.; Mizuno, S.; Torii, K. PIV measurement of internal structure of diesel fuel spray. *Exp. Fluids* **2000**, *29*, S211–S219.
196. Zhu, J.; Kuti, O.A.; Nishida, K. An investigation of the effects of fuel injection pressure, ambient gas density and nozzle hole diameter on surrounding gas flow of a single diesel spray by the laser-induced fluorescence-particle image velocimetry technique. *Int. J. Engine Res.* **2012**, *14*, 630–645.
197. Zhang, M.; Xu, M.; Hung, D.L.S. Simultaneous two-phase flow measurement of spray mixing process by means of high-speed two-color PIV. *Meas. Sci. Technol.* **2014**, *25*, 095204.
198. Rulliere, C. (Ed.) *Femtosecond Laser Pulses*; Springer: Berlin, Germany, 2005.

applied
sciences

MDPI

Article

In Situ Measurement of Alkali Metals in an MSW Incinerator Using a Spontaneous Emission Spectrum

Weijie Yan [1,*], Chun Lou [2,*], Qiang Cheng [2], Peitao Zhao [1] and Xiangyu Zhang [3]

[1] School of Electrical and Power Engineering, China University of Mining and Technology, No. 1, Daxue Road, Xuzhou 221116, Jiangsu, China; p.zhao@cumt.edu.cn

[2] State Key Laboratory of Coal Combustion, Huazhong University of Science and Technology, No. 1037, Luoyu Road, Wuhan 430074, Hubei, China; chengqiang@mail.hust.edu.cn

[3] National Engineering Research Center of Clean Coal Combustion, Xi'an Thermal Power Research Institute Co., Ltd, No. 136, Xingqing Road, Xi'an 710032, Shanxi, China; zhangxiangyu@tpri.com.cn

* Correspondence: yanweijie@cumt.edu.cn (W.Y.); Lou_chun@sina.com (C.L.);
  Tel.: +86-516-8359-2000 (W.Y.); +86-27-8754-2417 (C.L.)

Academic Editor: Johannes Kiefer
Received: 24 December 2016; Accepted: 6 March 2017; Published: 9 March 2017

Abstract: This paper presents experimental investigations of the in situ diagnosis of the alkali metals in the municipal solid waste (MSW) flame of an industrial grade incinerator using flame emission spectroscopy. The spectral radiation intensities of the MSW flame were obtained using a spectrometer. A linear polynomial fitting method is proposed to uncouple the continuous spectrum and the characteristic line. Based on spectra processing and a non-gray emissivity model, the flame temperature, emissivity, and intensities of the emission of alkali metals were calculated by means of measuring the spectral radiation intensities of the MSW flame. Experimental results indicate that the MSW flame contains alkali metals, including Na, K, and even Rb, and it demonstrates non-gray characteristics in a wavelength range from 500 nm to 900 nm. Peak intensities of the emission of the alkali metals were found to increase when the primary air was high, and the measured temperature varied in the same way as the primary air. The temperature and peak intensities of the lines of emission of the alkali metals may be used to adjust the primary airflow and to manage the feeding of the MSW to control the alkali metals in the MSW flame. It was found that the peak intensity of the K emission line had a linear relationship with the peak intensity of the Na emission line; this correlation may be attributed to their similar physicochemical characteristics in the MSW. The variation trend of the emissivity of the MSW flame and the oxygen content in the flue gas were almost opposite because the increased oxygen content suppressed soot formation and decreased soot emissivity. These results prove that the flame emission spectroscopy technique is feasible for monitoring combustion in the MSW incinerator in situ.

Keywords: MSW incineration; flame emission spectroscopy; flame temperature; in-situ measurement

1. Introduction

The rapid development of modern society has resulted in the production of an increasing volume of MSW. The disposal, treatment and management of MSW are common problems in the world. Incineration has been widely used for the disposal of MSW in order to conserve fuel and protect the environment by lessening the amount of MSW in landfills [1,2]. However, emissions of trace organic compounds—particularly polychlorinated dioxins and furans, and heavy metals such as mercury, lead and cadmium—are major environmental hazards caused by MSW incineration. Alkali and alkaline earth metals are typically found in high concentrations in MSW. These alkali metal compounds often contribute to the fouling and corrosion of the heating surfaces of fluidized bed and grate incinerators. To understand the transfer of heavy metals and alkali and alkaline earth metals, and to control the emission of these metals in MSW incineration, certain methods for measuring the metal species in flue gasses have been applied—including absorption, emission, plasma spectroscopy, and ionization/mass spectrometric methods [3,4]. In addition, other analysis techniques, such as X-ray fluorescence (XRF), energy dispersive X-ray spectroscopy (EDX), scanning electron microscope combined energy dispersive spectrometer (SEM-EDS) and X-ray diffraction (XRD), have been used to provide supporting/complementary data about the metal species in fuel and ash [5]. However, few in situ measuring systems focus on combustion or flame inside the incinerator. Only two of the studies we reviewed [6,7] used infrared thermographic cameras to replace thermocouples and to obtain a temperature map inside the combustion chambers of incineration plants. These studies demonstrated that in situ monitoring combustion can be helpful, particularly for control and optimization purposes. Some research has reviewed diagnostic techniques for monitoring and controlling flames [8,9]. Optical methods based on laser diagnostic techniques have been applied for measuring velocity, temperature, species, and particulate in flames. However, using these latter methods presents some challenges in practical industrial furnaces and incinerators, because of the limited optical access, the laser attenuation by the particulate medium and other factors. Flame emission spectroscopy (FES) is another alternative non-intrusive approach for diagnosing flames. The spontaneous emission spectra of flames contain continuous spectra from the blackbody radiation of solid particles (e.g., soot, char, or ash particles); from the band spectra produced by the radiation of gas molecules (e.g., CO_2 or H_2O); from the chemiluminescence of excited radicals (e.g., OH, CH, or C_2); and from the line spectra of free atoms (e.g., alkalis or alkaline earth metals) [10]. Based on the measurement and analysis of the spontaneous emission spectra of flames, FES has been proposed to determine equivalence ratios [9], to monitor flame stoichiometry [9,11] and to calculate flame temperatures in a natural gas-fired furnace [12], a pulverized coal-fired furnace [13,14], a gasification facility [15], an aluminum flame [16], and even an MSW incinerator [17]. For the flame of the MSW incinerator, strong characteristic emission lines of Na and K are observed in the visible spectra in the experiments, and more studies are needed to know if this type spectrum could be used for measurement temperature, while related works are relatively few in number.

In this paper, experimental investigations involving in situ measuring and analyzing spontaneous emission spectra of MSW flames in a grate incinerator using FES are presented. First, the experimental setup, including the grate incinerator and a spectrometer system, will be introduced, and then the measurement principle will be described. Then, the results of temperature, emissivity and intensities of the emissions lines of the alkali metals derived from MSW flame spectra in the incinerator will be given. The effects of operating parameters of the incinerator, such as the primary air on the measurement results, will be analyzed. Finally, some concluding remarks will be made.

2. Experimental Setup

Due to the high moisture and low heat value characteristics of Chinese MSW, the preferred furnace type for MSW incineration is a mechanical grate incinerator; the reciprocating grate incinerator is a particularly common choice, because it can decrease the pollutant emissions by means of a forward and backward pushing two-stage grate. Therefore, our experiment was conducted in a two-stage reciprocating grate incinerator of the GCL Renewable Energy Power Generation Co., Ltd., which is located in Xuzhou, China. The incinerator can dispose of MSW at a rate of 400 tons per day. The rated evaporative capacity of the incinerator is 35 ton/h. The MSW studied was a mixture of household trash (70%), street waste (20%), and office waste (10%). During the experiment, the MSW was sampled and analyzed three times. Table 1 lists the proximate and ultimate analysis of the MSW used in the experiment. The composition of the incinerated MSW incineration changed only slightly.

Table 1. Proximate and ultimate analysis of the MSW (air dried basis, %).

No.	Moisture	Ash	Volatile	Fixed Carbon	Total Sulfur	Hydrogen	Calorific Value
1	3.62	55.62	29.64	11.12	0.32	2.50	6.40 MJ/kg
2	3.59	54.63	29.72	12.06	0.41	2.60	6.63 MJ/kg
3	3.68	55.46	28.82	12.04	0.38	2.46	6.66 MJ/kg

As shown in Figure 1, the L-shaped furnace was adapted by the two-stage reciprocating grate waste incinerator to prolong the time that the flue gas would reside in the furnace and to prevent a flue gas blockade. The total length and width of the grate were 9.7 m and 7.07 m, respectively. The furnace outlet was located above the grate center, and the waste underwent drying, pyrolysis, and combustion progressively along the grate after entering the incinerator. To achieve the staged combustion process, the total air supply was divided into primary and secondary air. The primary air used for assisting combustion was heated to 230 °C in two stages of preheating by steam and flue gas, respectively. The residence time of the flue gas in the furnace was > 2 s at a temperatures > 850 °C after the secondary air supply was provided.

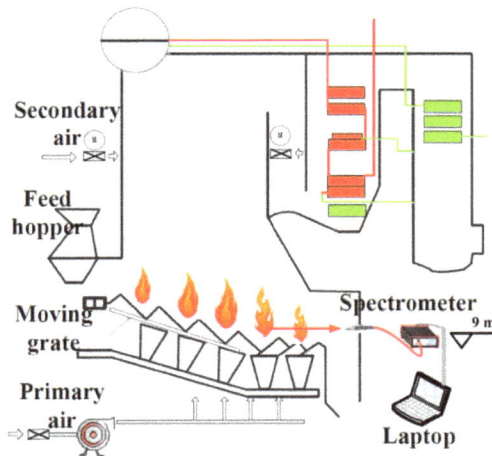

Figure 1. Schematic diagram of the experimental setup.

As shown in Figure 1, a spectrometer system, consisting of a portable spectrometer (AvaSpec-USB2048, Avantes, Apeldoorn, The Netherlands) with 2048 pixels, a fiber with a collimating lens, and a laptop was placed at the rear section of the combustion chamber. The measurement wavelength range of the spectrometer was 200 nm to 1100 nm, and the spectral resolution was 0.8 nm. In the process of spectral analysis, a smoothing function was introduced to decrease noise, each pixel will be averaged with three left and three right neighboring pixels, in this case, and the resolution of spectrum data is about 1.71 nm. The spectra of the MSW flames were collected by the fiber and then converted into digital signals by the spectrometer; finally, they were sent to the laptop via a USB cable. Dedicated application software developed by the authors of this study was used to obtain and process the MSW flame emission spectra in situ.

3. Measurement Principle

The spectra of the spontaneous emissions of flames with a certain wavelength range can be obtained by an emission spectrometer. Generally, the raw output of the spectrometer will be in photon counts $S(\lambda)$ of the radiation as a function of the wavelength λ; these counts represent the relative spectral profile of the flame. To quantitatively analyze the spectra of the flame's spontaneous emission, it is necessary to calibrate the profile of the spectral radiation intensities $I(\lambda)$ along the wavelength. As described in [14,15,17], a blackbody furnace was used to obtain the calibration coefficients $k(\lambda) = I_b(\lambda)/S_b(\lambda)$. Figure 2 shows the profiles of $k(\lambda)$ within the wavelength range from 500 nm to 900 nm at four temperatures: 1273, 1373, 1473, and 1573 K. It can be seen that profiles of the calibration coefficients at different blackbody temperatures were consistent with each other; thus, it can be concluded that the calibration coefficients remained independent of temperature. The flame's spectral radiation intensities can be obtained by $I(\lambda) = S(\lambda) \cdot k(\lambda)$.

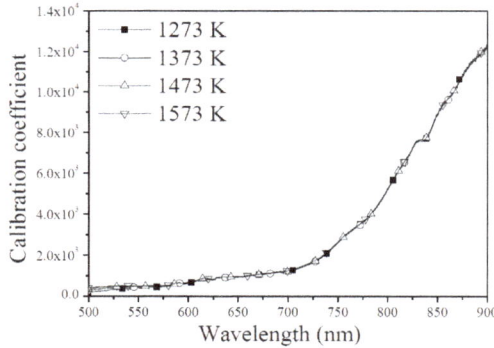

Figure 2. Calibration coefficients for the spectrometer at different blackbody temperatures.

Three cases of relative spectral profiles and their corresponding spectral radiation intensities of the MSW flame are shown in Figure 3. As shown in Figure 3b, the flame's spectral radiation intensities I_e were equal to the sum of the continuous spectral radiation intensities I_c produced by particulates in the flame plus the discontinuous spectral intensities I_d, such as the intensities of the emission lines of the alkali metals:

$$I_e = I_c + I_d \qquad (1)$$

where every term is wavelength dependent.

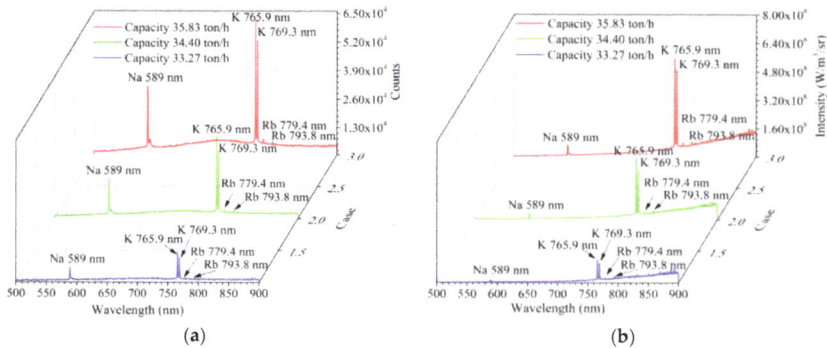

Figure 3. Three typical of MSW flame spectra. (**a**) Relative spectral profiles $S(\lambda)$; and (**b**) corresponding spectral radiation intensities $I(\lambda)$.

Some algorithms have been proposed to automatically separate the continuous and discontinuous spectral information from the measured flame emission spectra [18]. In this paper, continuous spectral radiation intensities in a wavelength range from 500 nm to 900 nm were extracted from the measured spectral radiation intensities of the MSW flame, and they were then fitted by a fourth-order polynomial. Figure 4 shows a comparison of the original continuous spectral radiation intensities (dotted line) extracted from Figure 3b and the corresponding fitting curves (solid line). In order to evaluate the fitting performance, the goodness-of-fit coefficient (GFC) [18] quality metrics between the fitting I_c' and the original I_c spectral radiation intensities in the absence of discontinuous emission lines were calculated as:

$$GFC = \frac{\left| \sum_j I_c(\lambda_j) I_c'(\lambda_j) \right|}{\sum_j \left[\left[I_c(\lambda_j) \right]^2 \right]^{1/2} \sum_j \left[\left[I_c'(\lambda_j) \right]^2 \right]^{1/2}} \tag{2}$$

Generally, an accurate fitting yields a GFC higher than 0.99. The results of the GFC for the three cases in Figure 4 were 0.9933, 0.9973, and 0.9988, respectively.

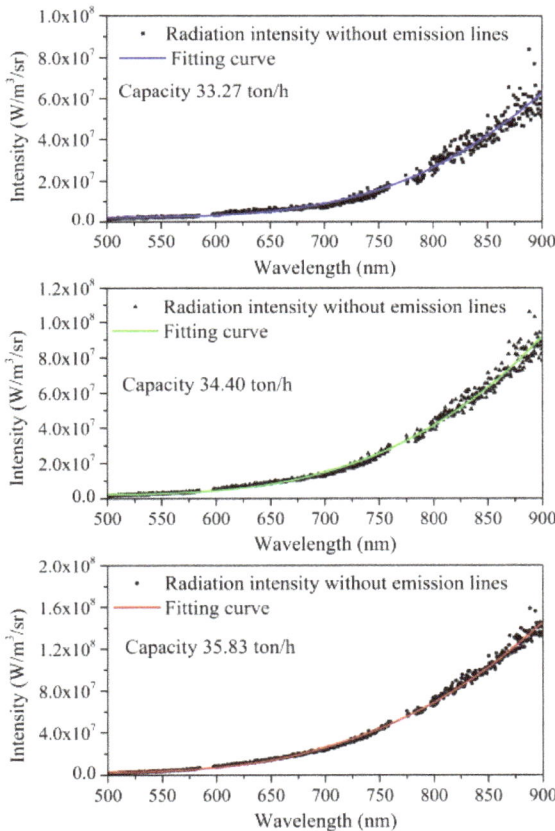

Figure 4. Comparison of the original continuous spectral radiation intensities (dotted line) separated from Figure 3b and the corresponding fitting curves (solid line).

The continuous spectral radiation intensities from the MSW flame may be described by the approximate form of Planck's radiation law for a heat radiation source at temperature T:

$$I_c(\lambda_i, T) = \varepsilon(\lambda_i) \cdot C_1 \lambda_i^{-5} \exp(-C_2/(\lambda_i T))/\pi \tag{3}$$

where $I_c(\lambda_i, T)$ is the radiation intensity at wavelength λ_i ($i = 1, 2, \cdots, m$). For the type of spectrometer used in this article, $m = 703$, T is the flame temperature, $\varepsilon(\lambda_i)$ is the emissivity at wavelength λ_i, and C_1 and C_2 are the first and second Plank's constants. For the MSW incineration, the pyrolysis of the MSW generated volatile matter, which was burned in the incinerator, producing more soot particles in the MSW flame. The soot emissivity is spectrally dependent; Hottel and Broughton [19,20] developed an empirical emissivity model for soot particles that has been widely used:

$$\varepsilon(KL, \lambda_i) = 1 - e^{-KL/\lambda_i^{\alpha}} \tag{4}$$

where $\varepsilon(KL, \lambda_i)$ is the spectral emissivity, KL is the optical thickness of the flame, and α is an empirical parameter depending upon the given wavelength, given by Hottel and Broughton as 1.39 in the visible spectral region. When substituted into Equation (3), this model allows one to calculate the temperature T and KL [19–21] numerically by using the least squares method. The calculated value of KL, which is independent of wavelength, can then be used with Equation (4) to determine the spectrally-dependent $\varepsilon(\lambda_i)$.

After T and $\varepsilon(\lambda_i)$ were obtained, the continuous spectral radiation intensities I_c were calculated using Equation (3). Then, the intensities of the emission lines of the alkali metals were calculated by $I_d = I_e - I_c$, which was used to analyze the variation of the peak intensities of the alkali metals' lines in the MSW flame.

4. Results and Analysis

4.1. MSW Flame Spectra

Figure 3 shows three examples of the spontaneous emission spectra of the MSW flame in the incinerator from 500 nm to 900 nm. As shown in Figure 3a, some emission peaks or lines at wavelengths of 589, 765.9, 769.3, 779.4, and 793.8 nm were observed. According to the National Institute of Standards and Technology (NIST)'s Atomic Spectra Database, these emission peaks represent the atomic emission lines of Na (588.995 nm, 589.592 nm), K (766.490 nm, 769.896 nm), and Rb (780.027 nm, 794.760 nm), respectively. Since the spectral resolution of this spectrometer was low, only 0.8 nm, the two emission lines of Na at approximately 589 nm were not distinguished. Generally, vegetable and garden wastes form a significant part of household trash, so the composition of the MSW contains Na and K. However, the presence of Rb in the MSW is seldom reported in the literature. It may come from small electronic components in the MSW. As the absolute spectral intensity distribution of the MSW flames depicted in Figure 3b shows, it can be seen that the peak intensities of the Rb emission lines are too weak to be identified. Continuous emission is mainly due to thermal radiation from soot particles in the MSW flame.

4.2. Temperature, Emissivity, and Alkaline Metal Emission Lines from MSW Flame Spectra

According to measurement principles, the temperature and emissivity can be derived from continuous spectral radiation intensities based on Hotel and Broughton's emissivity model. The emissivity distributions of the three MSW flame spectra represented in Figure 3b are given in Figure 5. These figures demonstrate that the MSW flame shows non-gray characteristics in a wavelength range of 500 nm to 900 nm, caused by an increased amount of soot particles in the flame. The corresponding temperatures of the three cases are 1282 K, 1335 K, and 1379 K, respectively.

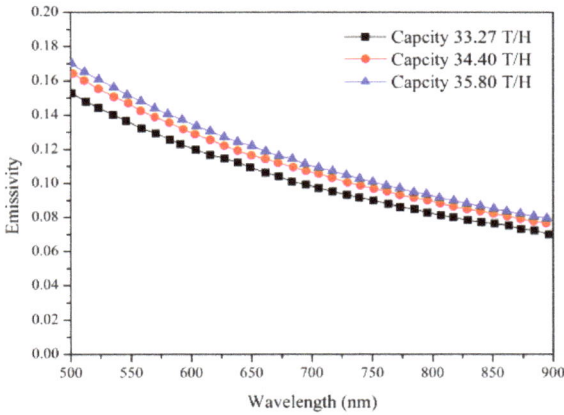

Figure 5. Emissivity distributions calculated from spectral radiation intensities in Figure 3b.

To validate the accuracy of the results, the spectral radiation intensities were re-calculated by substituting the values of the temperature and emissivity into Equation (3). Comparing the re-calculated and measured spectral radiation intensities, as shown in Figure 6, it was found that they matched well in the continuous wavelength range.

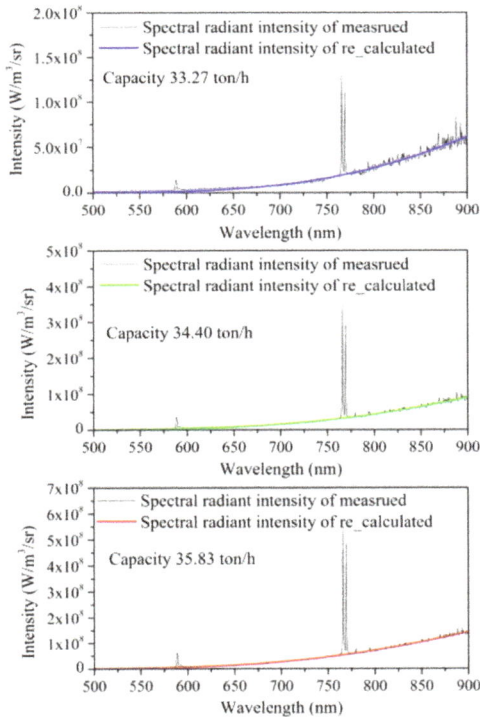

Figure 6. Comparison of the re-calculated and measured spectral radiation intensities.

As mentioned in Section 3, the intensities of the emission lines of the alkali metals were obtained by deducting the continuous spectral radiation intensities from the measured spectral radiation intensities. Figure 7 lists the intensities of the alkali metal emission lines of the three MSW flame spectra depicted in Figure 3b. The intensities of the alkali metal emission lines were taken to compare with those of the background radiation, and the results show that the relationship between them is nearly five times, so uncoupling the continuous spectrum and the characteristic lines is necessary for the temperature measurement of K and Na.

The strength of the alkali metals' emission lines is a complex function of the concentration of the alkali metals, and it also depends on the self-absorption of the radiation. For small-scale flames enriched by alkali metals, the concentration distributions of the alkali metals are generally assumed to be uniform. Some methods similar to those described in the literature [22] may be used to correct the effect of self-absorption. However, for an industrial MSW incinerator, which has a large furnace cavity, these concentration distributions of the alkali metals in the furnace are inappropriately assumed to be uniform. In such cases, it is difficult to correct for the effect of self-absorption using methods such as those described in the literature [23]. It is best to reconstruct the 2D and/or 3D distribution of local quantities inside the isothermal and homogeneous combustion chambers, as is often done for a large-scale, coal-fired boiler furnace by means of a visible flame image processing system with many detectors [24]. In this article, by means of a single spectrometer detector, line-of-sight values were determined for the measurement of the flame temperature, the radiation intensity of the alkali metals, and the flame emissivity. In this article, the peak intensities of the alkali metals' emission lines are used for in situ monitoring of the alkali metal content in the MSW flame.

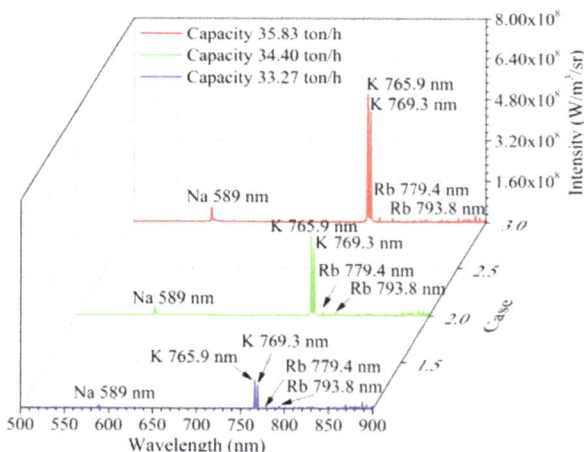

Figure 7. Intensities of alkali metals' emission lines calculated from spectral radiation intensities in Figure 3b.

4.3. Comparison of Measured Results and Operating Parameters of the Incinerator

Generally, in order to decrease the pollutant emissions, the incinerator is operated with an excess air coefficient of between 1.4 and 1.6, and at a temperature higher than 850 °C. The adjustment of the primary air is crucial for controlling the excess air and the temperature in MSW incineration. A test for changing the flow rate of the primary air flow from 28,000 to 35,000 m^3/h has been conducted. The measured temperature, emissivity, and peak intensities of the alkali metals' emission lines were compared with the recorded primary air, actual evaporative capacity, and oxygen content of the incinerator, respectively. Figure 8 displays the normalized peak intensities of Na (589 nm), K (765.9 nm), and Rb (779.4 nm) emission lines, and demonstrates that the flow rates of primary air varied with time. From Figure 8, it can be seen that when the primary air was high, the peak intensities of the alkali metal emissions increased. The increased primary air enhanced the particulate entrainment in the MSW flame, so more alkali metal entered the gaseous medium of the flame instead of staying in the ash. Therefore, the concentrations of alkali metals in the flame increased, which caused the peak intensities of the alkali metals' emission lines to increase. Changes in the peak intensities of the alkali metals' emission lines may be related to changes in the characteristics of the MSW. However, in this experiment, the composition of the MSW used changed only slightly.

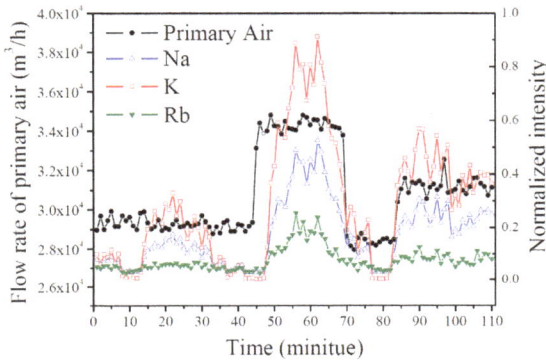

Figure 8. Peak intensities of the alkali metals' emission lines and primary air varying with time.

Additionally, the adjustment of the primary air affected the temperature of the MSW flame. Figure 9 displays temperatures derived from the MSW flame spectra and the primary air and shows that they varied with time. The temperature had a similar variation trend as the primary air. The increase in surface temperature on the particles leads to higher K and Na emissions; those results are similar to the release of alkali metal during biomass and coal combustion. The literature [24] shows that the release of K and Na from well-characterized biofuels was quantified as a function of temperature in a lab-scale fixed-bed reactor. Wang et al. [25] investigated the release and transformation fundamentals of sodium during pyrolysis of two Zhundong bituminous coals, with experimental results showing that the release of sodium increases with temperature, indicating that temperature has a profound effect on the volatilization of sodium. The range of temperatures was between 1200 and 1400 K. In this temperature range, the emission of pollutants, such as dioxin, can be controlled.

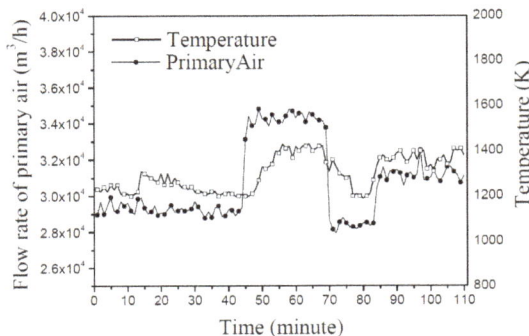

Figure 9. Temperatures of the MSW flame and primary air varying with time.

Another interesting phenomenon was that the peak intensity of the K emission line had a linear relationship with the peak intensity of the Na emission line, as shown in Figure 10. This correlation may be attributed to the wide availability of these two metals in the waste, and also to their similar physicochemical characteristics. Na and K are both important elements in biomass, and because biomass (vegetable, plant, and garden wastes) forms a significant part of household waste, it stabilizes the ratio of Na and K in the MSW.

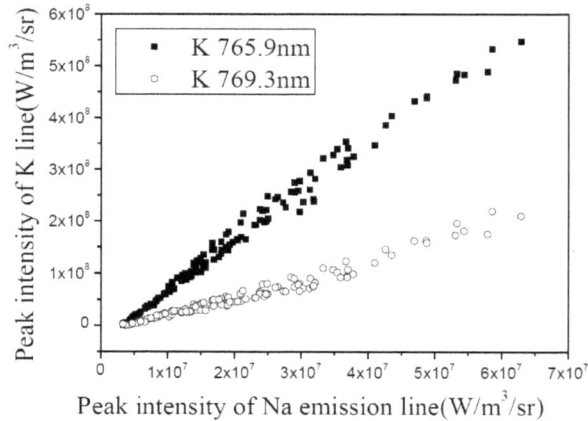

Figure 10. Relationship between peak intensities of K and Na emission lines.

Figure 11 displays the emissivity derived from the MSW flame spectra and oxygen content in the flue gas varying with time. It is interesting that the two parameters behaved in an approximately contrary manner, which may be explained by the soot in the flame. The increased oxygen content may have suppressed soot formation. Thus, emissivity from the soot particles generated in the MSW in the furnace may increase as the flame decreases.

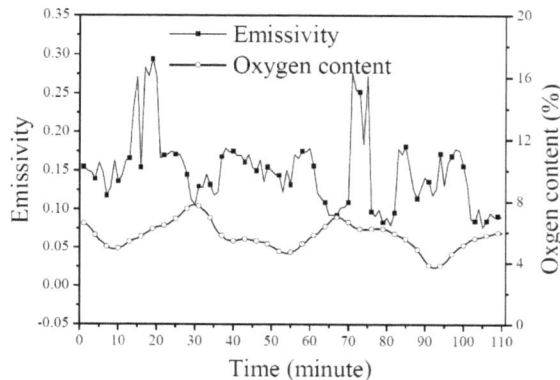

Figure 11. Emissivity and oxygen content varying with time.

5. Conclusions

In this paper, a flame emission spectroscopy technique was used to diagnose, in situ, the spontaneous emission spectra of MSW flames in an industrial-grade incinerator. For MSW flames, strong characteristic emission lines of Na and K are observed in the visible spectra, and a linear polynomial fitting method is proposed to uncouple them. Based on spectra processing and a non-gray emissivity model, the temperature, emissivity, and intensities of the emission lines of alkali metals were calculated from the spectral radiation intensities of the MSW flame. Experimental results indicate that the MSW flame contained alkali metals, including Na, K, and even Rb, and that the flame demonstrated non-gray characteristics in a wavelength range from 500 nm to 900 nm. The variation of the measured temperature, emissivity, and peak intensities of the alkali metals' emission lines may reflect the variation of actual oxygen content and primary air. The preliminary results prove that the flame emission spectroscopy technique is feasible for in situ monitoring of the combustion in a MSW incinerator. In later studies, calibration methods similar to the literature [22] will be conducted, making use of multi-parameter coupling technology to correlate the concentration of alkali metals, the flame temperature, and the radiation of alkali metals. Then the radiation function of alkali metal concentration will be obtained.

Acknowledgments: The present study has been supported by the Fundamental Research Funds for the Central Universities, CUMT: 2015QNA14, the National Natural Science Foundation of China (No. 51676077, No. 51676078).

Author Contributions: Weijie Yan and Chun Lou conceived and designed the experiments; Weijie Yan, Chun Lou, Peitao Zhao, and Xiangyu Zhang performed the experiments; Weijie Yan, Chun Lou, and Qiang Cheng analyzed the data; Weijie Yan and Chun Lou wrote the paper.

Conflicts of Interest: The founding sponsors had no role in the design of the study, in the collection, analyses, or interpretation of data, in the writing of the manuscript, and in the decision to publish the results.

References

1. Ruth, L.A. Energy from municipal solid waste: A comparison with coal combustion technology. *Prog. Energy Combust. Sci.* **1998**, *24*, 545–564. [CrossRef]
2. Zhao, X.G.; Jiang, G.W.; Li, A.; Li, Y. Technology, cost, a performance of waste-to-energy incineration industry in China. *Renew. Sustain. Energy Rev.* **2016**, *55*, 115–130.
3. Poole, D.; Sharifi, V.; Swithenbank, J.; Argent, B.; Ardelt, D. On-line detection of metal pollutant spikes in MSW incinerator flue gases prior to clean-up. *Waste Manag.* **2007**, *27*, 519–532. [CrossRef] [PubMed]
4. Monkhouse, P. On-line spectroscopic and spectrometric methods for the determination of metal species in industrial processes. *Prog. Energy Combust. Sci.* **2011**, *37*, 125–171. [CrossRef]
5. Vassilev, S.V.; Vassileva, C.G. Methods for characterization of composition of fly ashes from coal-fired power stations: A critical overview. *Energy Fuels* **2005**, *19*, 1084–1098. [CrossRef]
6. Schuler, F.; Rampp, F.; Martin, J.; Wolfrum, J. TACCOS—A thermography-assisted combustion control system for waste incinerators. *Combust. Flame* **1994**, *99*, 431–439. [CrossRef]
7. Manca, D.; Rovaglio, M. Infrared thermographic image processing for the operation and control of heterogeneous combustion chambers. *Combust. Flame* **2002**, *130*, 277–297. [CrossRef]
8. Kohse-Höinghaus, K.; Barlow, R.S.; Aldén, M.; Wolfrum, J. Combustion at the focus: Laser diagnostics and control. *Proc. Combust. Inst.* **2005**, *30*, 89–123. [CrossRef]
9. Ballester, J.; García-Armingol, T. Diagnostic techniques for the monitoring and control of practical flames. *Prog. Energy Combust. Sci.* **2010**, *36*, 375–411. [CrossRef]
10. Broida, H.P. The spectroscopy of flames. *Combust. Flame* **1957**, *1*, 487–488. [CrossRef]
11. Romero, C.; Li, X.; Keyvan, S.; Rossow, R. Spectrometer-based combustion monitoring for flame stoichiometry and temperature control. *Appl. Therm. Eng.* **2005**, *25*, 659–676. [CrossRef]
12. Keyvan, S.; Rossow, R.; Romero, C. Blackbody-based calibration for temperature calculations in the visible and near-IR spectral ranges using a spectrometer. *Fuel* **2006**, *85*, 796–802. [CrossRef]
13. Cai, X.S.; Cheng, Z.H.; Wang, S.M. Flame measurement and combustion diagnoses with spectrum analysis. *AIP Conf. Proc.* **2007**, *914*, 60.
14. Sun, Y.P.; Lou, C.; Zhou, H.C. A simple judgment method of gray property of flames based on spectral analysis and the two-color method for measurements of temperatures and emissivity. *Proc. Combust. Inst.* **2011**, *33*, 735–741. [CrossRef]
15. Parameswaran, T.; Hughes, R.; Gogolek, P.; Hughes, P. Gasification temperature measurement with flame emission spectroscopy. *Fuel* **2014**, *134*, 579–587. [CrossRef]

16. Christian, G.P.; Alexander, C.W.; David, M.S.; Donaldson, A.B.; Jonathan, L.H. Aluminum Flame Temperature Measurements in Solid Propellant Combustion. *Appl. Spectrosc.* **2014**, *68*, 362–366.

17. Yan, W.J.; Zhou, H.C.; Jiang, Z.W.; Lou, C.; Zhang, X.K.; Chen, D.L. Experiments on measurement of temperature and emissivity of municipal solid waste (MSW) combustion by spectral analysis and image processing in visible spectrum. *Energy Fuels* **2013**, *27*, 6754–6762. [CrossRef]

18. Arias, L.; Sbarbaro, D.; Torres, S. Removing baseline flame's spectrum by using advanced recovering spectrum techniques. *Appl. Opt.* **2012**, *51*, 6111–6116. [CrossRef] [PubMed]

19. Draper, T.S.; Zeltner, D.; Tree, D.R.; Xue, Y.; Tsiava, R. Two-dimensional flame temperature and emissivity measurements of pulverized oxy-coal flames. *Appl. Energy* **2012**, *95*, 38–44. [CrossRef]

20. Lu, G.; Yan, Y.; Riley, G.; Bheemul, H.C. Concurrent measurement of temperature and soot concentration of pulverized coal flames. *IEEE Trans. Instrum. Meas.* **2002**, *51*, 990–995.

21. Zhao, H.; Ladommatos, N. Optical diagnostics for soot and temperature measurement in diesel engines. *Prog. Energy Combust. Sci.* **1998**, *24*, 221–255. [CrossRef]

22. He, Y.; Zhu, J.J.; Li, B.; Wang, Z.H.; Li, Z.S.; Aldén, M.; Cen, K.F. In-situ measurement of sodium and potassium release during oxy-fuel combustion of lignite using laser-induced breakdown spectroscopy: Effects of O_2 and CO_2 concentration. *Energy Fuels* **2013**, *27*, 1123–1130. [CrossRef]

23. Wang, H.J.; Huang, Z.F.; Wang, D.D.; Luo, Z.X.; Sun, Y.P.; Fang, Q.Y.; Lou, C.; Zhou, H.C. Measurements on flame temperature and its 3D distribution in a 660 MW archfired coal combustion furnace by visible image processing and verification by using an infrared pyrometer. *Meas. Sci. Technol.* **2009**, *20*, 114006.

24. Frandsen, F.J.; Lith, S.C.; Korbee, R.; Yrjas, P.; Backman, R.; Obernberger, I.; Brunner, T.; Jöller, M. Quantification of the release of inorganic elements from biofuels. *Fuel Process. Technol.* **2007**, *88*, 1118–1128. [CrossRef]

25. Wang, C.A.; Jin, X.; Wang, Y.K.; Yan, Y.; Cui, J.; Liu, Y.H.; Che, D.F. Release and transformation of sodium during pyrolysis of Zhundong Coals. *Energy Fuels* **2015**, *29*, 78–85. [CrossRef]

applied
sciences

MDPI

Article

Collinear FAST CARS for Chemical Mapping of Gases

Anton Shutov [1,*], Dmitry Pestov [1,†], Narangerel Altangerel [1], Zhenhuan Yi [1], Xi Wang [1], Alexei V. Sokolov [1,2] and Marlan O. Scully [1,2,3]

[1]   Institute for Quantum Science and Engineering, Texas A&M University, College Station, TX 77843-4242, USA; dspestov@gmail.com (D.P.); nara.altan@gmail.com (N.A.); yzh@tamu.edu (Z.Y.); xwangphy@gmail.com (X.W.); sokol@physics.tamu.edu (A.V.S.); scully@tamu.edu (M.O.S.)

[2]   Quantum Optics Laboratory at the Baylor Research and Innovation Collaborative, Baylor University, Waco, TX 76706, USA

[3]   Department of Mechanical and Aerospace Engineering, Princeton University, Princeton, NJ 08544, USA

*   Correspondence: tony_shutov@physics.tamu.edu; Tel.: +1-979-204-9103

†   Current address: IPG Photonics Corporation, 377 Simarano Dr., Marlborough, MA 01752, USA.

Academic Editor: Johannes Kiefer
Received: 12 May 2017; Accepted: 2 July 2017; Published: 8 July 2017

Featured Application: remote spectroscopic mapping of gas emissions and leaks.

Abstract: We examine the concentration dependence of the Coherent Anti-Stokes Raman Scattering (CARS) signal obtained for gas mixtures at various conditions using the Femtosecond Adaptive Spectroscopic Technique (FAST). We use the CARS signal of the Q-branch vibrational oscillation of molecular oxygen (1556 cm^{-1}) to confirm the quadratic dependence of the coherent signal on the number of molecules in a test volume. In addition, we demonstrate multi-shot FAST CARS imaging of a gas flow in free space by raster-scanning the area of interest.

Keywords: optics; coherent anti-Stokes Raman scattering (CARS); Raman scattering; gas imaging; concentration dependence

1. Introduction

Both spontaneous and coherent Raman scattering effects have been employed in spectroscopic systems applied to trace detection and identification of chemicals [1,2] and as a method to measure concentrations of various components in gases [3–5], liquids and solids [6,7]. In spontaneous Raman scattering, the signal is directly proportional to the number of molecules interacting with a single beam of input light. By contrast, coherent anti-Stokes Raman scattering (CARS) is a nonlinear process where the molecules are first put into a state of coherent oscillations, resulting in an increased probability for a probe pulse to scatter and produce Stokes or anti-Stokes shifted photons [8]. Therefore, CARS-based techniques possess an inherent ability to offer higher signal strength and faster collection speed compared to approaches based on spontaneous Raman scattering [9].

With the development of ultrashort pulsed laser sources, CARS imaging opened the possibility to study the dynamics of such rapidly changing systems as gas flows [10]. The great flexibility of the CARS technique makes it a popular instrument for performing thermometry measurements for various gases: nitrogen and oxygen [11–13], carbon dioxide [14], as well as methane and hydrogen [15]. CARS techniques allow concentration and temperature measurements in harsh environments and at high temperature during the combustion process [16,17]. Moreover, one-dimensional [18,19], as well as two-dimensional [20] single-shot thermometry and mapping of reactants and products in gas flows can be performed by different experimental approaches.

The CARS signal is expected to scale proportionally to the second power of the concentration of target molecules in a sample; however, Raman linewidth changes [21], signal re-absorption [7], as well as nonresonant contribution from background molecules [22,23] may alter the dependence of the CARS signal on the concentration, making it difficult to correlate the signal strength with the distribution of those molecules. Gas flows, in particular, are multi-component systems, usually containing a large number of background molecules contributing to the CARS signal background. Some of the aforementioned problems, i.e., Raman line broadening, can be resolved by taking into account line broadening coefficients [24] as well as dephasing rates for the gas species [25].

Here we show that the femtosecond adaptive spectroscopic technique (FAST) for CARS [26] maintains the proper dependence of the CARS signal on concentration and thereby assures a direct correspondence between CARS intensity image and molecular distribution. FAST CARS is a combination of methods aimed at optimizing the CARS signal and suppress the noise background. In the 'hybrid' implementation of FAST CARS, we use two ultrashort pump-Stokes excitation pulses to prepare a coherent oscillation of molecules (Figure 1a), in combination with a narrowband delayed probe pulse to provide near-perfect non-resonant four-wave mixing background suppression [27,28]. Recently, in addition to gas-phase studies, this scheme was successfully used for microspectroscopy [29,30], and for spectroscopic molecular sensing aided by plasmonic nanostructures [31–33].

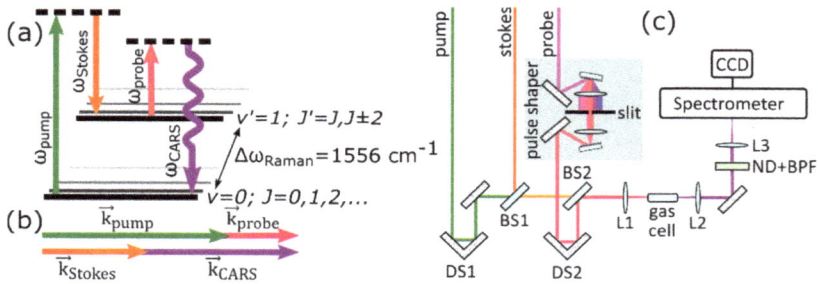

Figure 1. (**a**) CARS (coherent anti-stokes Raman scattering) level diagram. CARS signal is generated from the probe pulse scattering off the molecular vibration, coherently prepared by the pump (ω_{pump}) and Stokes (ω_{Stokes}) pulses, which are resonant with the Raman frequency of the molecule ($\Delta\omega_{Raman}$). v, J, v', J'—initial and final vibrational and rotational states correspondingly; (**b**) Phase matching CARS scheme for collinear beams configuration; (**c**) Experimental setup. DS1,2—delay stages. BS1,2—beamsplitters, L1–3—lenses (f = 100 mm), ND + BPF—set of neutral density and bandpass filters.

In this work, we experimentally demonstrate that the hybrid CARS signal obtained for molecular oxygen scales as the square of the number of molecules in two scenarios. We should note that the determination of molecular concentration based on the quadratic dependence of the CARS signal intensity on the number of molecules has been performed before using ns-CARS [34] and hybrid CARS [35] techniques. However, here we focus on demonstrating that this dependence can be maintained at different experimental conditions without prior calibration of specific gases composition or concentration. In addition, we provide a simple yet clear illustration of how the hybrid CARS technique can be exploited for imaging and sensing of a gas flow escaping from a nozzle by performing multi-shot measurements along with a raster-scanning an area in a collinear configuration. In the future, the same ideas can be applied to remote detection and sensing of trace amounts of gases [36].

2. Experimental Setup

Figure 1c shows the experimental setup for our oxygen concentration measurements. As a laser source, we employ a Ti:sapphire regenerative amplifier (1 kHz repetition rate, 0.73 mJ/pulse, Legend, Coherent, Santa Clara, CA, USA) with two equally pumped optical parametric amplifiers (OPAs: OPerA-VIS/UV and OPerA-SFG/UV Coherent, Santa Clara, CA, USA). The outputs of the two OPAs are used as the pump and Stokes pulses (<130 fs). A small fraction of the amplifier output passes through a homemade pulse shaper with an adjustable slit and is used as a probe beam. The adjustable slit width allows us to select a narrow spectral band ($\Delta v \approx 11$ cm^{-1}) at 806 nm, which corresponds to about 2 ps pulse duration. The pump and the probe beams pass through delay stages (DS1,2), so that the probe pulse is time delayed with respect to the pump and Stokes pulses. Choosing time delay for the probe pulse along with its pulse shaping, provide a clear CARS signal with suppressed non-resonant background from the four-wave mixing signal generated by the three beams.

The collinear geometry configuration for all beams aids to simplify optical alignment and signal collection, with automatically satisfied phase-matching conditions in gases (Figure 1b). The wavelength (FWHM bandwidth) of the pump, Stokes pulses and probe beams are set to 556 nm (13.6 nm), 610 nm (14.1 nm) and 806 nm (0.7 nm), with 3.1 µJ, 2.6 µJ and 0.9 µJ pulse energies respectively. When the probe delay is varied, the full hybrid CARS spectrogram can be observed (for example, see the graphical abstract for this paper). In present experiments, the probe delay with respect to the pump/Stokes pulses is fixed at an optimum values of 2.1 ps. The beams are combined by two dichroic beamsplitters (BS1,2) and focused by a convex lens L1 ($f = 100$ mm) in the homemade gas cell (Borosilicate glass, 1 mm thickness, 25.2 cm^3, 8 cm long) filled with N$_2$ and air gas mixture. The three beams and the generated CARS signal are collimated by another lens (L2, $f = 100$ mm) and filtered by a set of neutral density (FW2AND Thorlabs, Newton, NJ, USA) and bandpass filters (FF01-732/68 Semrock, Rochester, NY, USA) (ND+BPF). After passing through the filters, the beam is focused by the lens L3 ($f = 100$ mm) on the entrance slit of the spectrometer: a spectrograph (Chromex Spectrograph 250is, Albuquerque, NM, USA), which has a liquid-nitrogen-cooled CCD camera (CCD: uncoated Spec-10:400B, Princeton Instruments, Trenton, NJ, USA) attached.

The spectral resolution of the experimental setup is limited by two factors: resolution of the spectrometer and the probe pulse spectral width. The spectrometer resolution is determined by its slit size and was set to 0.12 cm^{-1} for all measurements, thus the total spectral resolution during experiments was mainly limited by the width of the probe pulse (≈ 11 cm^{-1}). The probe pulse duration (≈ 2 ps) constrains the temporal resolution. The spatial resolution in transverse to the beams propagation direction is mainly limited by a beam focal spot diameter (estimated at ≈ 20 µm). However, the axial spatial resolution, i.e., in the direction of the beams propagation, can be potentially limited by the fact that the CARS signal is generated in the region up to 6 times larger than the Rayleigh range (≈ 0.4 mm for our beams configuration) [8] in the collinear beams configuration. This fact does not limit the concentration dependence measurements since the gas cell length is significantly larger than the estimated length of this region (2.4 mm).

3. Results and Discussion

In the first set of experiments we study the CARS signal of O$_2$ molecules by filling the gas cell with air at different pressures. As the first step towards optimizing the beam alignment and time delays for pump/probe pulses, we collect the CARS spectrum showing rotational-vibrational structure of oxygen molecule (Figure 2), where several peaks can be clearly distinguished. The signal was collected for 2.2 s at atmospheric pressure at 1 kHz. The main peak of the spectrum (1555.6 cm^{-1}) corresponds to the Q-branch ($\Delta J = 0$) vibrational transition from the vibrational ground level of the oxygen molecule. Smaller peaks on the left and on the right to the Q-branch transition represent O- ($\Delta J = -2$) and S- ($\Delta J = +2$) transitions correspondingly [37].

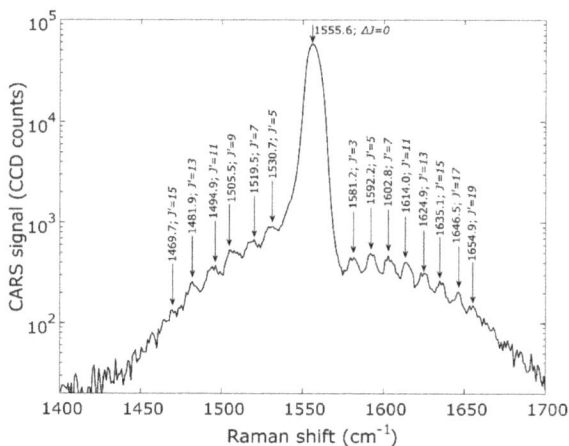

Figure 2. CARS spectrum of molecular oxygen in ambient air. Arrows mark the locations with maximum intensities, and corresponding wavenumbers and final quantum rotational numbers are given.

Thereafter, we use the maximum CCD count value from the region of Raman shift $1556 \pm 1 \text{ cm}^{-1}$, i.e., the intensity of the Q-branch vibrational line to examine the dependence of the CARS signal on gas concentration and pressure. Before calculating the maximum intensity, the zero-pressure background signal was subtracted from the data. We find using the maximum intensity to be more convenient and yet sufficiently precise for studying the dependence of the CARS signal on molecular concentration, when the absolute concentration of gas species is irrelevant. However, it is worth mentioning, that the fitting curves parameters for integrated intensity data points differ from the peak fittings by less than one standard deviation (4.6%).

During these measurements laser power fluctuations at 1 kHz repetition rate were below 0.5% and the signal was integrated for 5.2 s. The long integration time aids in significantly reducing the signal intensity fluctuations, but forces us to use at least ND = 3.0 filter at ambient conditions in order to avoid saturation of the CCD.

First, the cell is filled with a gas mixture of ambient air and pure nitrogen (Figure 3, solid circles). We assume the oxygen presence in air to be 21% and vary the partial pressures of air and nitrogen while keeping the total pressure in the cell constant. These measurements allow us to study the dependence of CARS signal on the concentration of O_2 with different amounts of background molecules (N_2). In this case, the CARS signal scales as a square of the relative oxygen concentration.

Next, we fill the cell with ambient air at various pressures, keeping the gas composition and ratio of oxygen molecules to background molecules constant. We control the total pressure inside the gas cell by a ball valve, and perform measurements for the range from 0.01 bar to 1.07 bar. The experimental data with a fit curve are shown in Figure 3 (hollow diamonds), where the signal is proportional to the square of the gas pressure. Hence, we conclude that in both cases of constant gas mixture at different pressures and varied gas compositions at constant pressure, the CARS signal is proportional to the square of the number of O_2 molecules.

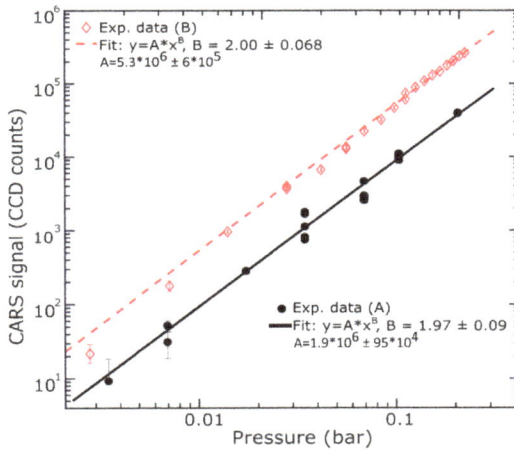

Figure 3. CARS signal dependence on O_2 partial pressure at constant total pressure in the cell (black circles, solid line), and at constant gas mixture at different pressures in the cell (red diamonds, dashed line). Both fittings are performed using "power1" fit in MATLAB R2016b (MathWorks, Natick, MA, USA). 95% confidence bounds are provided for each fitting parameter.

In the next set of experiments, we replace the gas cell with a 1 mm round nozzle connected to a cylinder with N_2 gas and pointed in the direction perpendicular to the beams propagation. Hereafter, we obtain the CARS signal for O_2 molecules in ambient air in front of the nozzle in the vicinity of the focal plane of the beam (Figure 4), where zero of the X-axis corresponds to the nozzle surface and zero of the Y-axis to the center of the nozzle. After we set the gas pressure such that the gas flow from the nozzle remains constant (estimated at ≈4.5 m/s), we move the nozzle in a direction transverse to the axis of the beam propagation direction (Y and X axes in the figure). Hence, by obtaining the CARS signal for O_2 at various nozzle positions it becomes possible to visualize the flow of nitrogen from the nozzle. One can see that the nitrogen flow stays almost symmetric about the zero of the Y-axis as it propagates away from the nozzle, where the slight slope can be due to a tilt of the nozzle. The CARS signal from oxygen in the central part of the flow increases very slowly, and the nitrogen diffuses into the surrounding air.

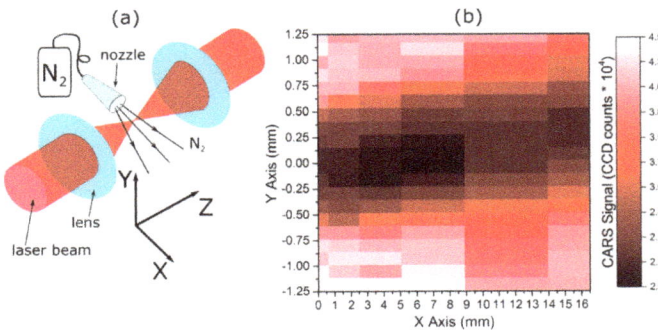

Figure 4. N_2 flow as it displaces air: (**a**) Setup schematics; (**b**) the CARS signal from O_2 in front of the nozzle. Darker regions correspond to higher concentrations of nitrogen.

We then add a thin metal plate (50 × 20 × 0.5 mm) in front of the nozzle to examine the resulting flow disturbance (Figure 5). The plate is placed at a distance 5.9 mm away from the nozzle surface in such a manner that the top half of the nozzle is blocked by the plate. The thin plate acts as an impenetrable barrier/obstacle for the gas and laser beams. The flow is disturbed and no longer symmetric; points with zero CARS signal mark the plate location. Moreover, the signal decreases in front of the plate as the nitrogen flow is partially redirected along the surface, while another portion of the flow is deflected by the plate downward. Right behind the barrier, the signal is restored to its value in ambient air since the nitrogen flow cannot penetrate through the plate; i.e., the air in this region stays undisturbed.

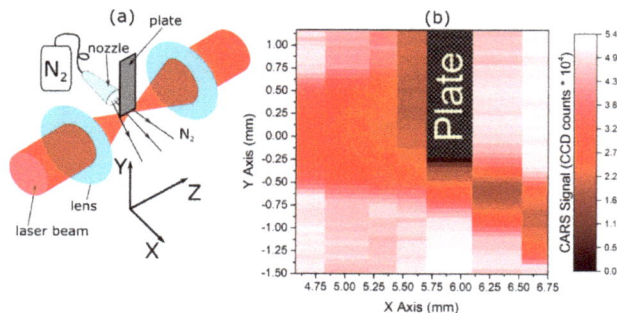

Figure 5. N_2 flow as it displaces air with a flat barrier plate placed in front of the nozzle. (**a**) Setup schematic; (**b**) O_2 CARS signal from air in front of the nozzle. Darker regions correspond to higher concentrations of nitrogen.

4. Conclusions

We have experimentally demonstrated the quadratic dependence of the CARS signal of the Q-branch vibrational transition of molecular oxygen on the number of O_2 molecules in a gas mixture. We examined two cases. In the first, the mixture was prepared with various amount of background molecules but at constant total pressure. In the second case, the mixture composition remained unchanged while the pressure was varied. No significant difference between these cases was found as both signals showed quadratic dependence on the number of molecules. Furthermore, we have illustrated how CARS can be used for the visualization of gas flow in a simple, free-space configuration, both with a plate barrier and without it. We believe this method is applicable for performing gas flow images for any molecules with any Raman-active modes, as long as the CARS signal can be retrieved with a suppressed non-resonant background.

Acknowledgments: This works is supported by Office of Naval Research (Awards No. N00014-16-1-3054, N00014-16-1-2578), Robert A. Welch Foundation (Grants No. A-1261, A-1547), National Science Foundation (PHY-1307153 and CHE-1609608). Anton Shutov is supported by the Herman F. Heep and Minnie Belle Heep Texas A&M University Endowed Fund held/administered by the Texas A&M Foundation. We thank Alexandra Zhdanova and Mariia Shutova for proofreading the manuscript.

Author Contributions: Anton Shutov, Dmitry Pestov and Alexei V. Sokolov designed the experiment and analyzed the data. Anton Shutov, Dmitry Pestov, Narangerel Altangerel, Zhenhuan Yi, Xi Wang performed the experiment. Alexei V. Sokolov, Marlan O. Scully supervised the work. Anton Shutov wrote the manuscript with help of Alexei V. Sokolov, Dmitry Pestov and Zhenhuan Yi.

Conflicts of Interest: The authors declare no conflict of interest.

References

1. Kneipp, K.; Kneipp, H.; Itzkan, I.; Dasari, R.R.; Feld, M.S. Ultrasensitive Chemical Analysis by Raman Spectroscopy. *Chem. Rev.* **1999**, *99*, 2957–2976. [CrossRef] [PubMed]

2. Dogariu, A.; Goltsov, A.; Pestov, D.; Sokolov, A.V.; Scully, M.O. Real-time detection of bacterial spores using coherent anti-Stokes Raman spectroscopy. *J. Appl. Phys.* **2008**, *103*, 036103. [CrossRef]

3. Webber, B.F.; Long, M.B.; Chang, R.K. Two-dimensional average concentration measurements in a jet flow by Raman scattering. *Appl. Phys. Lett.* **1979**, *35*, 119–121. [CrossRef]

4. Regnier, P.R.; Moya, F.; Taran, J.P.E. Gas Concentration Measurement by Coherent Raman Anti-Stokes Scattering. *AIAA J.* **1974**, *12*, 826–831. [CrossRef]

5. Richardson, D.R.; Lucht, R.P.; Kulatilaka, W.D.; Roy, S.; Gord, J.R. Chirped-probe-pulse femtosecond coherent anti-Stokes Raman scattering concentration measurements. *JOSA B* **2013**, *30*, 188–196. [CrossRef]

6. Dogariu, A.; Goltsov, A.; Xia, H.; Scully, M.O. Concentration dependence in coherent Raman scattering. *J. Mod. Opt.* **2008**, *55*, 3255–3261. [CrossRef]

7. Zhi, M.; Pestov, D.; Wang, X.; Murawski, R.K.; Rostovtsev, Y.V.; Sariyanni, Z.E.; Sautenkov, V.A.; Kalugin, N.G.; Sokolov, A.V. Concentration dependence of femtosecond coherent anti-Stokes Raman scattering in the presence of strong absorption. *JOSA B* **2007**, *24*, 1181–1186. [CrossRef]

8. Nibler, J.W.; Knighten, G.V. Coherent Anti-Stokes Raman Spectroscopy. In *Raman Spectroscopy of Gases and Liquids*; Weber, P.D.A., Ed.; Topics in Current Physics; Springer: Berlin/Heidelberg, Germany, 1979; pp. 253–299. ISBN 978-3-642-81281-1.

9. Petrov, G.I.; Arora, R.; Yakovlev, V.V.; Wang, X.; Sokolov, A.V.; Scully, M.O. Comparison of coherent and spontaneous Raman microspectroscopies for noninvasive detection of single bacterial endospores. *Proc. Natl. Acad. Sci. USA* **2007**, *104*, 7776–7779. [CrossRef] [PubMed]

10. Roy, S.; Gord, J.R.; Patnaik, A.K. Recent advances in coherent anti-Stokes Raman scattering spectroscopy: Fundamental developments and applications in reacting flows. *Prog. Energy Combust. Sci.* **2010**, *36*, 280–306. [CrossRef]

11. Roy, S.; Kulatilaka, W.D.; Richardson, D.R.; Lucht, R.P.; Gord, J.R. Gas-phase single-shot thermometry at 1 kHz using fs-CARS spectroscopy. *Opt. Lett.* **2009**, *34*, 3857–3859. [CrossRef] [PubMed]

12. Reichardt, T.A.; Schrader, P.E.; Farrow, R.L. Comparison of gas temperatures measured by coherent anti-Stokes Raman spectroscopy (CARS) of O_2 and N_2. *Appl. Opt.* **2001**, *40*, 741–747. [CrossRef] [PubMed]

13. Matthäus, G.; Demmler, S.; Lebugle, M.; Küster, F.; Limpert, J.; Tünnermann, A.; Nolte, S.; Ackermann, R. Ultra-broadband two beam CARS using femtosecond laser pulses. *Vib. Spectrosc.* **2016**, *85*, 128–133. [CrossRef]

14. Kerstan, M.; Makos, I.; Nolte, S.; Tünnermann, A.; Ackermann, R. Two-beam femtosecond coherent anti-Stokes Raman scattering for thermometry on CO2. *Appl. Phys. Lett.* **2017**, *110*, 021116. [CrossRef]

15. Dedic, C.E.; Miller, J.D.; Meyer, T.R. Dual-pump vibrational/rotational femtosecond/picosecond coherent anti-Stokes Raman scattering temperature and species measurements. *Opt. Lett.* **2014**, *39*, 6608–6611. [CrossRef] [PubMed]

16. Braeuer, A.; Beyrau, F.; Weikl, M.C.; Seeger, T.; Kiefer, J.; Leipertz, A.; Holzwarth, A.; Soika, A. Investigation of the combustion process in an auxiliary heating system using dual-pump CARS. *J. Raman Spectrosc.* **2006**, *37*, 633–640. [CrossRef]

17. Tröger, J.W.; Meißner, C.; Seeger, T. High temperature O_2 vibrational CARS thermometry applied to a turbulent oxy-fuel combustion process: O_2 vibrational CARS thermometry for oxy-fuel combustion process. *J. Raman Spectrosc.* **2016**, *47*, 1149–1156. [CrossRef]

18. Kulatilaka, W.D.; Stauffer, H.U.; Gord, J.R.; Roy, S. One-dimensional single-shot thermometry in flames using femtosecond-CARS line imaging. *Opt. Lett.* **2011**, *36*, 4182–4184. [CrossRef] [PubMed]

19. Bohlin, A.; Kliewer, C.J. Direct Coherent Raman Temperature Imaging and Wideband Chemical Detection in a Hydrocarbon Flat Flame. *J. Phys. Chem. Lett.* **2015**, *6*, 643–649. [CrossRef] [PubMed]

20. Bohlin, A.; Kliewer, C.J. Single-shot hyperspectral coherent Raman planar imaging in the range 0–4200 cm^{-1}. *Appl. Phys. Lett.* **2014**, *105*. [CrossRef]

21. Roh, W.B.; Schreiber, P.W. Pressure dependence of integrated CARS power. *Appl. Opt.* **1978**, *17*, 1418–1424. [CrossRef] [PubMed]

22. Roy, S.; Meyer, T.R.; Gord, J.R. Time-resolved dynamics of resonant and nonresonant broadband picosecond coherent anti-Stokes Raman scattering signals. *Appl. Phys. Lett.* **2005**, *87*, 264103. [CrossRef]

23. Wang, X.; Zhang, A.; Zhi, M.; Sokolov, A.V.; Welch, G.R. Glucose concentration measured by the hybrid coherent anti-Stokes Raman-scattering technique. *Phys. Rev. A* **2010**, *81*, 013813. [CrossRef]

24. Millot, G.; Saint-Loup, R.; Santos, J.; Chaux, R.; Berger, H.; Bonamy, J. Collisional effects in the stimulated Raman Q branch of O_2 and O_2–N_2. *J. Chem. Phys.* **1992**, *96*, 961–971. [CrossRef]

25. Miller, J.D.; Roy, S.; Gord, J.R.; Meyer, T.R. Communication: Time-domain measurement of high-pressure N_2 and O_2 self-broadened linewidths using hybrid femtosecond/picosecond coherent anti-Stokes Raman scattering. *J. Chem. Phys.* **2011**, *135*, 201104. [CrossRef] [PubMed]

26. Scully, M.O.; Kattawar, G.W.; Lucht, R.P.; Opatrný, T.; Pilloff, H.; Rebane, A.; Sokolov, A.V.; Zubairy, M.S. FAST CARS: Engineering a laser spectroscopic technique for rapid identification of bacterial spores. *Proc. Natl. Acad. Sci. USA* **2002**, *99*, 10994–11001. [CrossRef] [PubMed]

27. Prince, B.D.; Chakraborty, A.; Prince, B.M.; Stauffer, H.U. Development of simultaneous frequency- and time-resolved coherent anti-Stokes Raman scattering for ultrafast detection of molecular Raman spectra. *J. Chem. Phys.* **2006**, *125*, 44502. [CrossRef] [PubMed]

28. Pestov, D.; Murawski, R.K.; Ariunbold, G.O.; Wang, X.; Zhi, M.; Sokolov, A.V.; Sautenkov, V.A.; Rostovtsev, Y.V.; Dogariu, A.; Huang, Y.; et al. Optimizing the Laser-Pulse Configuration for Coherent Raman Spectroscopy. *Science* **2007**, *316*, 265–268. [CrossRef] [PubMed]

29. Shen, Y.; Voronine, D.V.; Sokolov, A.V.; Scully, M.O. A versatile setup using femtosecond adaptive spectroscopic techniques for coherent anti-Stokes Raman scattering. *Rev. Sci. Instrum.* **2015**, *86*, 083107. [CrossRef] [PubMed]

30. Shen, Y.; Voronine, D.V.; Sokolov, A.V.; Scully, M.O. Single-beam heterodyne FAST CARS microscopy. *Opt. Express* **2016**, *24*, 21652–21662. [CrossRef] [PubMed]

31. Voronine, D.V.; Sinyukov, A.M.; Hua, X.; Wang, K.; Jha, P.K.; Munusamy, E.; Wheeler, S.E.; Welch, G.; Sokolov, A.V.; Scully, M.O. Time-Resolved Surface-Enhanced Coherent Sensing of Nanoscale Molecular Complexes. *Sci. Rep.* **2012**, *2*, 891. [CrossRef] [PubMed]

32. Ballmann, C.W.; Cao, B.; Sinyukov, A.M.; Sokolov, A.V.; Voronine, D.V. Dual-tip-enhanced ultrafast CARS nanoscopy. *New J. Phys.* **2014**, *16*, 083004. [CrossRef]

33. Hua, X.; Voronine, D.V.; Ballmann, C.W.; Sinyukov, A.M.; Sokolov, A.V.; Scully, M.O. Nature of surface-enhanced coherent Raman scattering. *Phys. Rev. A* **2014**, *89*, 043841. [CrossRef]

34. Beyrau, F.; Seeger, T.; Malarski, A.; Leipertz, A. Determination of temperatures and fuel/air ratios in an ethene–air flame by dual-pump CARS. *J. Raman Spectrosc.* **2003**, *34*, 946–951. [CrossRef]

35. Engel, S.R.; Miller, J.D.; Dedic, C.E.; Seeger, T.; Leipertz, A.; Meyer, T.R. Hybrid femtosecond/picosecond coherent anti-Stokes Raman scattering for high-speed CH4/N2 measurements in binary gas mixtures: Hybrid fs/ps CARS for high-speed CH4/N2 measurements. *J. Raman Spectrosc.* **2013**, *44*, 1336–1343. [CrossRef]

36. Hemmer, P.R.; Miles, R.B.; Polynkin, P.; Siebert, T.; Sokolov, A.V.; Sprangle, P.; Scully, M.O. Standoff spectroscopy via remote generation of a backward-propagating laser beam. *Proc. Natl. Acad. Sci. USA* **2011**, *108*, 3130–3134. [CrossRef] [PubMed]

37. Fletcher, W.H.; Rayside, J.S. High resolution vibrational Raman spectrum of oxygen. *J. Raman Spectrosc.* **1974**, *2*, 3–14. [CrossRef]

applied
sciences

MDPI

Article

Infrared Spectroscopy for Studying Structure and Aging Effects in Rhamnolipid Biosurfactants

Johannes Kiefer [1,2,3,*], Mohd Nazren Radzuan [4,5] and James Winterburn [4]

[1] Technische Thermodynamik and MAPEX Center for Materials and Processes, Universität Bremen, Badgasteiner Straße 1, 28215 Bremen, Germany

[2] School of Engineering, University of Aberdeen, Fraser Noble Building, Aberdeen AB24 3UE, UK

[3] Erlangen Graduate in Advanced Optical Technologies (SAOT), Friedrich-Alexander-Universität Erlangen-Nürnberg, 91054 Erlangen, Germany

[4] School of Chemical Engineering and Analytical Science, The University of Manchester, Manchester M13 9PL, UK; mohdnazren.radzuan@postgrad.manchester.ac.uk (M.N.R.); james.winterburn@manchester.ac.uk (J.W.)

[5] Department of Biological and Agricultural Engineering, Faculty of Engineering, Universiti Putra Malaysia, Serdang 43400, Malaysia

* Correspondence: jkiefer@uni-bremen.de; Tel.: +49-421-2186-4777

Academic Editor: Jun Kubota
Received: 23 April 2017; Accepted: 19 May 2017; Published: 22 May 2017

Abstract: Biosurfactants are produced by microorganisms and represent amphiphilic compounds with polar and non-polar moieties; hence they can be used to stabilize emulsions, e.g., in the cosmetic and food sectors. Their structure and its changes when exposed to light and elevated temperature are yet to be fully understood. In this study, we demonstrate that attenuated total reflection infrared (ATR-IR) spectroscopy is a useful tool for the analysis of biosurfactants, using rhamnolipids produced by fermentation as an example. A key feature is that the analytical method does not require sample preparation despite the high viscosity of the purified natural product.

Keywords: vibrational spectroscopy; FTIR; *Pseudomonas aeruginosa*; rhamnose; fatty acid

1. Introduction

Surface-active compounds that are produced by microorganisms are commonly referred to as biosurfactants [1]. Biosurfactants have been experiencing an ever-increasing interest in recent years because they have many potential commercial applications, e.g., in the food, personal care, and pharmaceutical sectors. Moreover, they have a number of advantageous properties including their low toxicity and biodegradability.

Rhamnolipids are a type of glycolipid biosurfactants. They consist of a glycosyl head group in terms of a rhamnose moiety and a fatty acid tail. Rhamnolipids are commonly produced by various strains of the microorganism *Pseudomonas aeruginosa* [2], a Gram-negative, rod-shaped bacterium. The type of rhamnolipid produced depends on the culture conditions, carbon source used, and the microbial strain [3]. The two main types of rhamnolipids are mono-rhamnolipids and di-rhamnolipids, which contain one or two rhamnose units, respectively.

The analysis of rhamnolipids is commonly done off-line using chromatographic and mass spectrometric methods [4,5]. These methods require sampling and, in addition, often sample preparation. Therefore, they are not suitable for in-line or on-line monitoring of the production process, or to do rapid screening of the final product. The fast analysis with minimal sample preparation is a major challenge here, but it could revolutionize the production and quality control of biosurfactants. Vibrational spectroscopy in terms of infrared (IR) and Raman spectroscopy can be a solution [6–8]. Especially IR spectroscopy, when performed in attenuated-total reflection (ATR) mode, seems to be a good option as it can be applied to opaque samples. In ATR-IR spectroscopy, the infrared radiation is propagating in a material with a refractive index (often referred to as internal reflection element, IRE) that is higher than that of the sample. Consequently, the radiation can undergo total internal reflection at the sample/IRE interface. Due to the light-matter-interaction in the evanescent field, the reflected radiation carries information about the absorption spectrum [9]. The penetration depth of the evanescent field in the sample is typically of the order of one micrometer [10,11].

To date, only a few studies reporting the application of vibrational spectroscopy to rhamnolipids have been published. Li et al. [12] applied IR spectroscopy together with a variety of analytical methods to a rhamnolipid-layered double hydroxide (RL-LDH) nanocomposite material. They aimed at a comprehensive characterization of the material and at unravelling how the rhamnolipid anions are introduced into the interlayers of LDH. ATR-IR spectroscopy for the qualitative identification of rhamnolipids produced in fermentation processes has been reported as well [13–19].

The present study aims at testing the potential of ATR-IR for the structural analysis and the investigation of aging effects of rhamnolipids. For this purpose, rhamnolipids have been produced and purified. IR spectroscopic analysis was performed after the purification process and after five months of storage, during which the samples were kept at room temperature.

2. Materials and Methods

The rhamnolipid samples were produced by *Pseudomonas aeruginosa* PAO1 which was supplied by Ulster University (Coleraine, Northern Ireland) from their collection. To purify the product, solid phase extraction (SPE) was used in normal phase mode using silica as the sorbent. A detailed description of the production and processing can be found in a recent article [20]. Five different runs resulted in five different samples. Due to the production in a biological process, the chemical composition of the samples varies. The samples were initially analyzed after the purification and a second time after keeping them at room temperature for five months. The second series of measurements was carried to test whether or not any signs of aging and deterioration can be observed in the spectra.

ATR-IR spectra were acquired on an Agilent Cary 630 (Agilent Technologies, Santa Clara, CA, USA) equipped with a diamond ATR unit (1 reflection, 45°). The range 4000–650 cm^{-1} was recorded with a nominal resolution of 2 cm^{-1}. To obtain an appropriate signal-to-noise ratio, 16 scans were averaged for each sample. The rhamnolipids had a high viscosity and are sticky. Prior to each measurement, the diamond crystal was cleaned using 2-propanol, ethanol, and acetone. Three measurements of individual samples taken from the five products obtained in different runs of the bioreactor were carried out to check the reproducibility and potential inhomogeneities. The differences observed were negligible and the spectra shown later represent the average of the three acquisitions.

3. Results

As natural products, rhamnolipids are a class of compounds rather than a pure substance. Key characteristics of a sample are the number of rhamnose and carboxylic acid/ester units. Figure 1 illustrates the structures of four common compounds contained in a rhamnolipid biosurfactant. The most common rhamnolipids comprise of one or two rhamnose units and one or two fatty acid chains with C10 or C12 chain length. The distinction between mono- and di-rhamnolipids is the most important task for an analytical tool [21,22].

Figure 2 displays the spectra of the five different samples obtained from different fermentation experiments. Note that the diamond crystal has very low transmission between 1900 and 2300 cm^{-1}; hence, the spectra exhibit experimental artifacts rather than meaningful signatures in this range. Most of the rest of the spectrum can be interpreted in a straightforward manner and Table 1 summarizes the observed signatures. The broad band between 3700 and 3100 cm^{-1} is due to O–H stretching vibrations. The range 3100–2700 cm^{-1} contains the C–H stretching modes. The fingerprint region, i.e., below 1750 cm^{-1}, is very rich in information. At 1707 cm^{-1} the C=O stretching mode of the carboxylic acid groups can be found. The high-wavenumber wing of this band shows a shoulder bands at

~1740 cm^{-1}, which can be attributed to C=O stretching of the ester groups. At lower wavenumbers, the C–H bending; C–C and C–O stretching; and a multitude of rocking, scissoring, wagging, and twisting modes can be found. A detailed assignment is rather complicated as all these modes can be overlapping with each other, and hence the assignment in Table 1 is a tentative one. However, as aforesaid, the main task is to distinguish between the different classes of compounds. Therefore, we focus on the differences of the spectra.

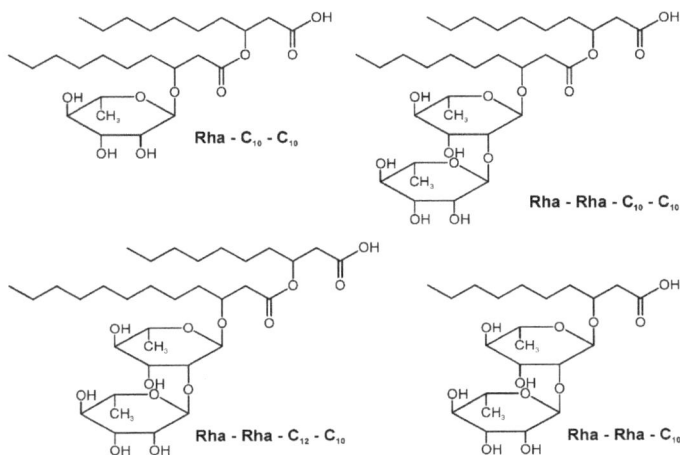

Figure 1. Chemical structures of four typical rhamnolipid compounds.

Table 1. Center wavenumbers of observed IR peaks in cm^{-1} and their tentative assignment. Some of the bands in the fingerprint region are not assigned due to possible ambiguity. s = shoulder, sym = symmetric, asym = anti-symmetric, str = stretching, bend = bending, sciss = scissoring, wag = wagging, rock = rocking.

R1	R2	R3	R4	R5	Assignment
		ν/cm^{-1}			
3274	3274	3288	3288	3285	OH str
2954s	2954s	2954s	2954s	2954s	CH$_3$ asym str
2926	2926	2926	2927s	2926	CH$_2$ asym str
-	-	-	2916	-	CH$_2$ asym str
2871s	2871s	2871s	2871s	2871s	CH$_3$ sym str
2855	2855	2855	2850	2855	CH$_2$ sym str
1740s	1740s	1740s	1740s	1740s	C=O str ester
1707	1707	1707	1707	1707	C=O str acid
1669	1669	1647	1647	1647	OH bend (residual water)
1592	1592	1590s	1591s	1591s	COO asym str
1527	1527	1527	1527	1527	Residual CHCl$_3$
1484	1484	1484s	-	-	Residual HCCl$_3$
1442	1442	1442	1442	1442	CH$_2$ sciss
1380	1380	1380s	1380s	1380s	COO sym str
1292	1292	1297s	1301s	1301	CH$_2$ wag, COH bend
1276s	1277	1277	1274	1273	CH$_2$ wag, COH bend
1241	1242	1242s	1242s	1242s	
1214	1214	1207	1206	1206	Residual CHCl$_3$
1155	1155	1155	1156	1156	CH rock
1120	1120	1121	1121	1121	CO str, CH rock
1032	1032	1031	1031	1031	CO str

Table 1. *Cont.*

R1	R2	R3	R4	R5	Assignment
		ν/cm^{-1}			
981	981	983	983	983	
964	964	965	965	966	
954s	954s	-	-	-	
918	918	917	917	918	
849	849	849	850s	850s	
818	818	818s	818s	818s	
809	809	809	809	809	
752	753	755	756	756	residual $CHCl_3$
-	735s	-	-	-	CH rock
724s	723s	724	724	724	CH rock
700	701	701	701	701	CH rock

The spectra of the five samples suggest that there are two groups of spectra exhibiting significant similarities: (1) R1 and R2; and (2) R3, R4, and R5. This visual observation has been confirmed by a principal component analysis. Four features highlighting spectral differences are marked by arrows in Figure 2. The most distinguishable signatures that are present only or at least significantly stronger in R1 and R2 can be found at 1592, 1484, 1380, 1241, and 850 cm^{-1}. The band at 1592 cm^{-1} is likely due to the anti-symmetric stretching of deprotonated carboxylate groups. While R1 and R2 exhibit a distinct peak, there are weak shoulder bands at this position. The same observation is made at 1380 cm^{-1}, which can be attributed to the corresponding symmetric mode. The other bands that appear differently for R1 and R2 and R3–5 are related with CH deformation vibrations and hence it is difficult to assign them unambiguously to specific structural differences without involving further analytical techniques.

Further differences between R1 and R2 and R3–5 are found at high wavenumber. The O–H stretching band around 3300 cm^{-1} is stronger in R3–5. The latter indicates that the R1 and R2 samples contain a lower number of rhamnose units. Probably, mono-rhamnolipids are dominating here while di-rhamnolipids are mainly present in R3–5. The presence of mainly di-rhamnolipids was also suggested by reference measurements with mass spectrometry, see Table 2. The enhanced IR signatures at around 1600 and 1240 cm^{-1} in the R1 and R2 samples can be attributed to ester groups indicating that these samples contain more fatty acids than R3–5. Hence, we can conclude that the IR spectrum is a suitable indicator of the structure and of the number of rhamnose units.

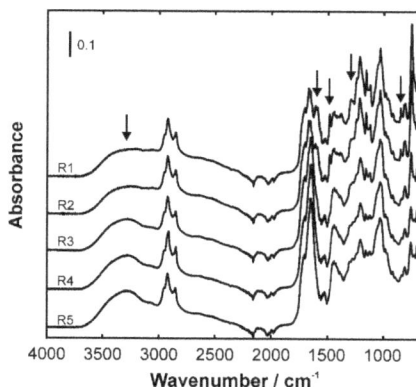

Figure 2. IR spectra of the five rhamnolipid samples. The arrows mark spectral position where significant changes can be observed.

Table 2. Main peaks from the MS analysis and assignment.

Rhamnolipid	*m/z* Value	Relative Intensity	Molecule Structure
Mono-rhamnolipid	503	15	Rha-C_{10}-C_{10}
Di-rhamnolipid	479	33	Rha-Rha-C_{10}
	649	100	Rha-Rha-C_{10}-C_{10}
	677	14	Rha-Rha-C_{12}-C_{10}

In order to analyze whether or not ATR-IR spectroscopy can be used to study aging effects as well, a second set of spectra was recorded after five months. During this period, the samples were kept at room temperature (298 K) but in darkness to avoid additional effects due to photo bleaching. Figure 3 shows the difference spectra between the initial and aged samples. Consequently, positive contributions mean that the signature reduced in absorbance over time; negative contributions mean the opposite.

As there are clear changes in the spectra, it can be concluded that aging takes place. On the other hand, there are only a few signatures in the difference spectra that are similar in all five cases, e.g., the positive double peak in the CH stretching region. The initial spectra exhibit four main peaks in this range: 2954, 2924, 2871, and 2854 cm^{-1}. The two peaks in the difference spectra are at 2915 and 2850 cm^{-1}. The 2954 and 2924 cm^{-1} peaks can be assigned to the anti-symmetric stretching modes of the CH_3 and CH_2 groups, respectively. The peaks at 2871 and 2854 cm^{-1} are due to the corresponding symmetric modes. The observed behavior indicates a decrease in the number of CH_2 groups. The dissociation of the fatty acid chains can explain this. Further support for the hypothesis that the fatty acid part of the rhamnolipids is mainly affected during the aging is provided by the negative contribution around 1700 cm^{-1}. In this region, the C=O stretching modes of the ester groups can be found. The spectra suggest that these groups are increasing during aging. This could be interpreted in terms of an ongoing esterification reaction. Unravelling the detailed mechanisms, however, would require further analysis and additional methods. This is beyond the scope of the present study.

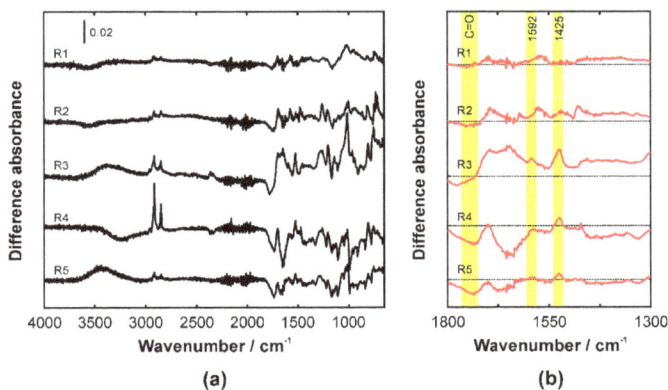

Figure 3. (a) Difference spectra of the initial and aged rhamnolipid samples; (b) zoomed in region between 1300 and 1800 cm^{-1}.

4. Conclusions

In this study, the application of infrared spectroscopy to rhamnolipid biosurfactants has been demonstrated. Five samples of rhamnolipids have been produced by fermentation and purified prior to the IR analysis. The IR spectra indicate that some samples are dominated by mono- and other by di-rhamnolipids. Moreover, measurements of the samples after five months of aging at room temperature indicate systematic changes in the chemical structure. These changes are mainly attributed to vibrational signatures of the fatty acid groups. In conclusion, our study shows that IR spectroscopy is capable of analyzing the chemical structure of biosurfactants and monitoring their aging. Future work in our labs will focus on systematic investigations of the aging effects to unravel the detailed molecular mechanisms. For this purpose, single compound reference samples will be analyzed during treatment with heat and ultraviolet radiation.

Acknowledgments: Johannes Kiefer gratefully acknowledges support from Deutsche Forschungsgemeinschaft (DFG) through grant KI1396/4-1. Mohd Nazren Radzuan gratefully acknowledges financial support from the Ministry of Education, Malaysia and the Department of Biological and Agricultural Engineering, Universiti Putra Malaysia under "Hadiah Skim Latihan IPTA (SLAI)". The authors wish to thank Ibrahim M. Banat at Ulster University for supplying the microbial strain used in this study and Reynard Spiess at the Manchester Institute of Biotechnology, University of Manchester for assistance with mass spectroscopy.

Author Contributions: Mohd Nazren Radzuan and James Winterburn conceived and designed the fermentation experiments; Mohd Nazren Radzuan performed the experiments; Johannes Kiefer conceived and designed the IR experiments; Johannes Kiefer performed the IR experiments; Johannes Kiefer analyzed the IR data. All authors contributed to data interpretation and preparation of manuscript with Johannes Kiefer putting together the first draft.

Conflicts of Interest: The authors declare no conflict of interest. The founding sponsors had no role in the design of the study; in the collection, analyses, or interpretation of data; in the writing of the manuscript, or in the decision to publish the results.

References

1. Abdel-Mawgoud, A.M.; Lépine, F.; Déziel, E. Rhamnolipids: Diversity of structures, microbial origins and roles. *Appl. Microbiol. Biotechnol.* **2010**, *86*, 1323–1336. [CrossRef] [PubMed]
2. Henkel, M.; Müller, M.M.; Kügler, J.H.; Lovaglio, R.B.; Contiero, J.; Syldatk, C.; Hausmann, R. Rhamnolipids as biosurfactants from renewable resources: Concepts for next-generation rhamnolipid production. *Process. Biochem.* **2012**, *47*, 1207–1219. [CrossRef]
3. Silva, S.N.R.L.; Farias, C.B.B.; Rufino, R.D.; Luna, J.M.; Sarubbo, L.A. Glycerol as substrate for the production of biosurfactant by *Pseudomonas aeruginosa* UCP0992. *Colloids Surf. B* **2010**, *79*, 174–183. [CrossRef] [PubMed]
4. Behrens, B.; Helmer, P.O.; Tiso, T.; Blank, L.M.; Hayen, H. Rhamnolipid biosurfactant analysis using on-line turbulent flow chromatography-liquid chromatography-tandem mass spectrometry. *J. Chromatogr. A* **2016**, *1465*, 90–97. [CrossRef] [PubMed]
5. Behrens, B.; Engelen, J.; Tiso, T.; Blank, L.M.; Hayen, H. Characterization of rhamnolipids by liquid chromatography/mass spectrometry after solid-phase extraction. *Anal. Bioanal. Chem.* **2016**, *408*, 2505–2514. [CrossRef] [PubMed]
6. Noack, K.; Eskofier, B.; Kiefer, J.; Dilk, C.; Bilow, G.; Schirmer, M.; Buchholz, R.; Leipertz, A. Combined shifted-excitation Raman difference spectroscopy and support vector regression for monitoring the algal production of complex polysaccharides. *Analyst* **2013**, *138*, 5639–5646. [CrossRef] [PubMed]
7. Koch, C.; Posch, A.E.; Herwig, C.; Lendl, B. Comparison of Fiber Optic and Conduit Attenuated Total Reflection (ATR) Fourier Transform Infrared (FT-IR) Setup for In-Line Fermentation Monitoring. *Appl. Spectrosc.* **2016**, *70*, 1965–1973. [CrossRef] [PubMed]
8. Zhao, L.; Fu, H.Y.; Zhou, W.C.; Hu, W.S. Advances in process monitoring tools for cell culture bioprocesses. *Eng. Life Sci.* **2015**, *15*, 459–468. [CrossRef]
9. Griffiths, P.R.; De Haseth, J.A. *Fourier Transform Infrared Spectrometry*, 2nd ed.; Wiley: New York, NY, USA, 2007.
10. Kiefer, J.; Frank, K.; Schuchmann, H.P. Attenuated total reflection infrared (ATR-IR) spectroscopy of a water-in-oil emulsion. *Appl. Spectrosc.* **2011**, *65*, 1024–1028. [CrossRef] [PubMed]
11. Averett, L.A.; Griffiths, P.R. Effective path length in attenuated total reflection spectroscopy. *Anal. Chem.* **2008**, *80*, 3045–3049. [CrossRef] [PubMed]
12. Li, Y.; Bi, H.Y.; Jin, Y.S. Facile preparation of rhamnolipid-layered double hydroxide nanocomposite for simultaneous adsorption of p-cresol and copper ions from water. *Chem. Eng. J.* **2017**, *308*, 78–88. [CrossRef]
13. Lahkar, J.; Borah, S.N.; Deka, S.; Ahmed, G. Biosurfactant of *Pseudomonas aeruginosa* JS29 against Alternaria solani: The causal organism of early blight of tomato. *Biocontrol* **2015**, *60*, 401–411. [CrossRef]

14. Antoniou, E.; Fodelianakis, S.; Korkakaki, E.; Kalogerakis, N. Biosurfactant production from marine hydrocarbon-degrading consortia and pure bacterial strains using crude oil as carbon source. *Front. Microbiol.* **2015**, *6*, 274. [CrossRef] [PubMed]

15. Zhao, F.; Zhang, J.; Shi, R.J.; Han, S.Q.; Ma, F.; Zhang, Y. Production of biosurfactant by a *Pseudomonas aeruginosa* isolate and its applicability to in situ microbial enhanced oil recovery under anoxic conditions. *RSC Adv.* **2015**, *5*, 36044–36050. [CrossRef]

16. Moussa, T.A.A.; Mohamed, M.S.; Samak, N. Production and characterization of di-rhamnolipid produced by *Pseudomonas aeruginosa* TMN. *Braz. J. Chem. Eng.* **2014**, *31*, 867–880. [CrossRef]

17. Singh, A.K.; Cameotra, S.S. Rhamnolipids Production by Multi-metal-Resistant and Plant-Growth-Promoting Rhizobacteria. *Appl. Biochem. Biotechnol.* **2013**, *170*, 1038–1056. [CrossRef] [PubMed]

18. Raheb, J.; Hajipour, M.J. The Characterization of Biosurfactant Production Related to Energy Consumption of Biodesulfurization in *Pseudomonas aeruginosa* ATCC9027. *Energy Sources A* **2012**, *34*, 1391–1399. [CrossRef]

19. Arutchelvi, J.; Doble, M. Characterization of glycolipid biosurfactant from *Pseudomonas aeruginosa* CPCL isolated from petroleum-contaminated soil. *Lett. Appl. Microbiol.* **2010**, *51*, 75–82. [CrossRef] [PubMed]

20. Radzuan, M.N.; Banat, I.M.; Winterburn, J. Production and characterization of rhamnolipid using palm oil agricultural refinery waste. *Bioresour. Technol.* **2017**, *225*, 99–105. [CrossRef] [PubMed]

21. Liu, J.F.; Wu, G.; Yang, S.Z.; Mu, B.Z. Structural characterization of rhamnolipid produced by Pseudonomas aeruginosa strain FIN2 isolated from oil reservoir water. *World J. Microbiol. Biotechnol.* **2014**, *30*, 1473–1484. [CrossRef] [PubMed]

22. George, S.; Jayachandran, K. Analysis of Rhamnolipid Biosurfactants Produced Through Submerged Fermentation Using Orange Fruit Peelings as Sole Carbon Source. *Appl. Biochem. Biotechnol.* **2009**, *158*, 694–705. [CrossRef] [PubMed]

*applied
sciences*

MDPI

Article

Distinguishing Different Cancerous Human Cells by Raman Spectroscopy Based on Discriminant Analysis Methods

Mingjie Tang [1], Liangping Xia [1,2,*], Dongshan Wei [1], Shihan Yan [1], Chunlei Du [1,*] and Hong-Liang Cui [1,3]

[1] Chongqing Key Laboratory of Multi-Scale Manufacturing Technology, Chongqing Institute of Green and Intelligent Technology, Chinese Academy of Sciences, Chongqing 400714, China; mjtang@cigit.ac.cn (M.T.); dswei@cigit.ac.cn (D.W.); yanshihan@cigit.ac.cn (S.Y.); hlcui2012@126.com (H.-L.C.)

[2] School of Electronic Information Engineering, Yangtze Normal University, Chongqing 408100, China

[3] College of Instrumentation Science and Electrical Engineering, Jilin University, Changchun 130061, China

* Correspondence: xialp@cigit.ac.cn (L.X.); cldu@cigit.ac.cn (C.D.);
  Tel.: +86-023-6593-5646 (L.X.); +86-023-6593-5600 (C.D.)

Received: 20 July 2017; Accepted: 30 August 2017; Published: 1 September 2017

Abstract: An approach to distinguish eight kinds of different human cells by Raman spectroscopy was proposed and demonstrated in this paper. Original spectra of suspension cells in the frequency range of 623~1783 cm^{-1} were acquired and pre-processed by baseline calibration, and principal component analysis (PCA) was employed to extract the useful spectral information. To develop a robust discrimination model, a linear discriminant analysis (LDA) and quadratic discriminant analysis (QDA) were attempted comparatively in the work. The results showed that the QDA model is better than the LDA model. The optimal QDA model was generated with 12 principal components. The classification rates are 100% in the calibration and prediction set, respectively. From the experimental results, it is concluded that Raman spectroscopy combined with appropriate discriminant analysis methods has significant potential in human cell detection.

Keywords: Raman spectra; discriminant analysis; distinguish; human cells

---

1. Introduction

Cancer is one of the main causes of human death in recent years [1]. Early diagnosis of cancer is a prerequisite for patient recovery [2], however the human body has many organs which may produce cancer cells, so there are many types of cancer cells. Therefore, the classification of cancer cells is also critical for the location of cancer incidence site. Not only that, another major feature of cancer is that it is prone to metastasis [3]. For example, the patient will bleed heavily undergoing tumor resection in the early stage of cancer, so cancerous cells may enter the peripheral blood circulation system, moving in the blood vessels in the form of a single cell or cell clusters, called circulating tumor cells. So, it is easy for cancer cells to migrate through the blood system. Therefore, the accurate identification of cancer cells is of great significance for diagnosing the metastasis, diffusion, and recurrence of cancer cells.

At present, the fluorescent labelling method is mainly used in the identification of the type of cancer cells due to its specificity. Fluorescent labelling is based on the specific binding of antigen and antibody [4]. The method has a substantial impact on, and even damage to, the original physiological activity of cells, which is not conducive to further analysis and research. It is prone to false positives for antigen and antibody specific binding [5]. In addition, the treatment of samples is complex, costly, and inefficient, so there are many drawbacks in clinical applications. If there is a non-contact technique which can specifically identify cancer cells at the physical level, it will not only keep the cell activity intact, but can also effectively solve the problem of efficiency in the complexity of biological sample pre-treatment. Raman spectroscopy is such a technique, which is a kind of inelastic scattering fingerprint spectra of molecules [6]. There is a strong specificity to reflect the changes in biochemical components of living cells in aqueous solutions without any labelling and fixation [7], as such Raman spectroscopy has been employed in clinical diagnostics, toxicology tests, and tissue engineering [8,9].

Raman spectroscopy is a fast, accurate, label-free, and non-destructive analytical tool for the detection of the human cells at the single cell level [10,11]. It can be used to obtain the difference of the intranuclear genetic material between the cancer cells and the normal cells, and the differences of the proteins in the cell membrane and cytoplasm [6,12]. It is known that cellular biochemical components vary depending upon the cancer cells coming from different organs, and different malignancy degrees. The difference is critical for the development of Raman spectroscopy as a new clinical diagnostic approach [10,11,13–15]. The main objective of the present experimental study was to investigate the biochemical difference in these different cancer cells (SH-SY5Y, HeLa, HO-8910, MDA-MB-231, U87, A549), the cells of distinct malignancy degree (MDA-MB-231 and MCF-7) and the normal cell line and cancer cells (HEB and U87) utilizing Raman spectroscopy.

In recent years, Raman spectroscopy combined with discriminant analysis techniques has drawn considerable attention for distinguishing similar biological materials such as tissues, cells, and biological molecules [16–22]. In this work, a rapid approach for distinguishing eight kinds of different human cells by Raman spectroscopy was studied. To develop an accurate Raman spectroscopic discrimination model, principal component analysis (PCA) was employed to extract useful spectral information, and then two discriminant analysis algorithms, linear discriminant analysis (LDA), and quadratic discriminant analysis (QDA) were employed to and contrasted to discriminate the eight different human cells.

2. Materials and Methods

2.1. Sample Preparation

All human cell samples belong to the eight different cell types, the name and the serial number can be seen in Table 1. Each cell type was divided into two groups at random. 2/3 of the samples were regarded as the calibration set and 1/3 of the samples were regarded as the prediction set. Dulbecco's Modified Eagle Medium (DMEM) was used to culture the eight different human cells, it was added 1% penicillin–streptomycin and 10% fetal bovine serum (both from Invitrogen, Grand Island, NY, USA), the cells were cultured at 37 °C with 5% CO_2 in a humidified atmosphere. Cells at a density of 1×10^6 per 1 mL of media were cultured on 25 cm^2 flask for around 24 h prior to experiments. Figure 1 shows an optical image of the morphology of the eight different adherent human cells. Before the Raman spectroscopy measurement, the cells were removed with 0.25% Trypsin-EDTA and then harvested in 3 mL PBS.

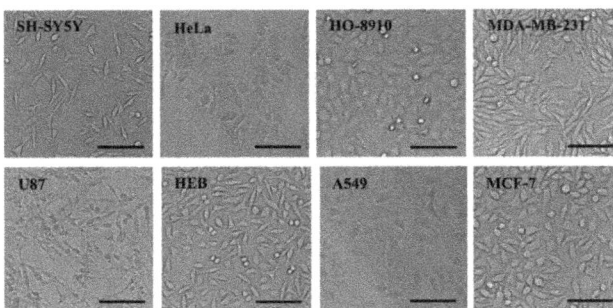

Figure 1. Optical images of the eight different human cells. Scale bar: 100 μm.

Table 1. Summary of the eight different human cells.

Sample's Serial Number	Sample's Name
1–100	SH-SY5Y, human neuroblastoma cells
101–200	HeLa, human cervical cancer cells
201–300	HO-8910, human ovarian cancer cells
301–400	MDA-MB-231, human breast cancer cells
401–500	U87, human glioma cells
501–600	HEB, human glioma cell line
601–700	A549, human lung cancer cells
701–800	MCF-7, human breast cancer cells

2.2. Raman Spectroscopy Measurement

A Renishaw inVia Raman spectrometer (controlled by WiRE 3.4 software, Renishaw plc, Wotton-under-Edge, UK) was used to collect the Raman spectra of the eight different human cells. It was connected to a Leica microscope (Leica DMLM, Leica Microsystems, Buffalo Grove, IL, USA), and equipped with a 532 nm laser that was focused through a 50×, NA = 0.75 objective (Leica Microsystems, Buffalo Grove, IL, USA); A standard calibration peak of 520.5 ± 0.1 cm^{-1} was used for the system with a silicon in a static mode. 20 µL cell suspension was dripped onto MgF$_2$ wafer for Raman spectrum measurement. The Raman spectra ranged from 623 to 1783 cm^{-1} were collected at 10 s laser exposure for 1 accumulation in a static mode. The laser power is 0.5 mW, BLZ of a diffraction grating is 2400 line/mm. The three replicate measurements at different times were performed for each cell to reduce the measurement error. The humidity and the temperature was kept at a stable level in the laboratory. Figure 2a presents the raw Raman spectra of the background and A549 cell samples.

For the Raman spectral pre-processing, Renishaw WiRE 3.4 software (Renishaw plc, Wotton-under-Edge, UK) was used to remove the cosmic rays in the raw spectra. Then, all of the Raman spectra were baseline corrected using the Vancouver Raman algorithm [23]. The smoothing pre-treatments were performed to reduce the external noises, and enhance the useful information of the biochemical composition. Therefore, Figure 2b gives representative peaks of the A549 cells after pre-processing. The average spectra of the eight different human cells after pre-processing are presented in Figure 3.

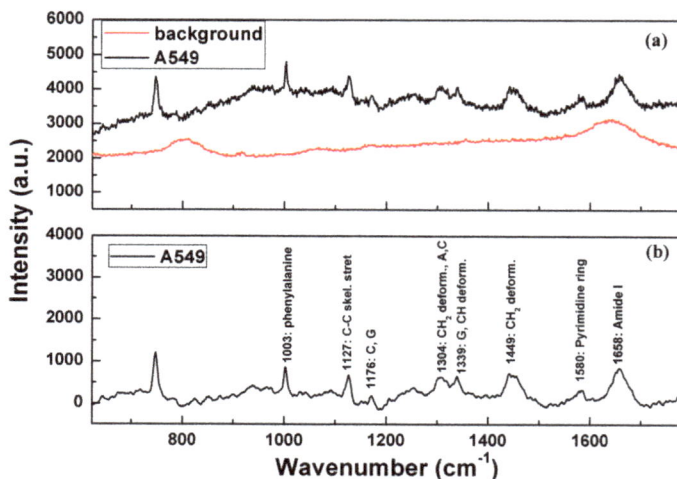

Figure 2. (**a**) Raman spectra of the background and A549 human cells; (**b**) The representative peaks of the A549 cells after pre-processing.

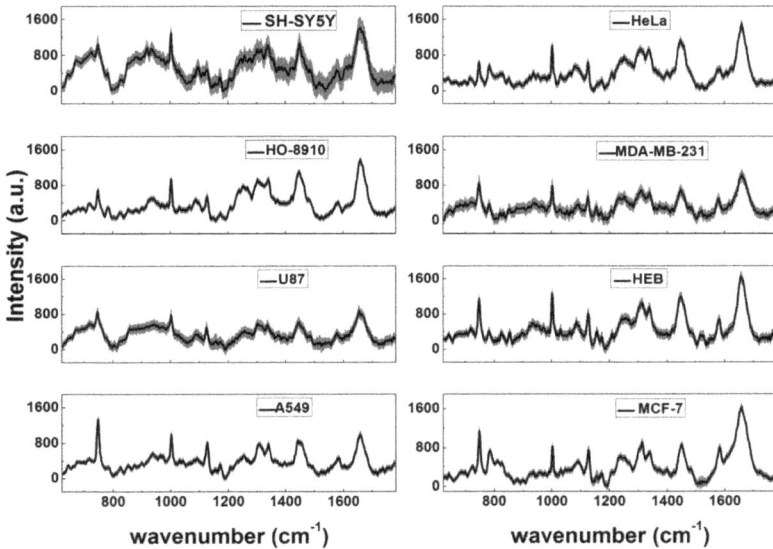

Figure 3. The average Raman spectra of the eight different human cells subtracted background obtained from baseline calibration. Error bars are standard deviation of the mean.

2.3. Software

All the analysis algorithms were executed in Matlab R2009a (MathWorks, Inc. Natick, MA, USA) under Windows XP.

3. Results and Discussion

3.1. Principal Component Analysis (PCA)

Raman spectral data were the application of the array of 1005 variables which were cross-sensitive toward the different biochemical composition in the human cells, so they contained overlapped information which brought some difficulty to the study. Multivariate data analysis could be applied to solve the problem. It is possible that PCA can extract the main information from the Raman spectra, and eliminate some of the overlapped information [7,24,25]. In the work, PCA based on spectra pre-processing method was the first attempted to visualize and extract the useful information from the multivariate spectral data to examine the qualitative differences among all types of cell samples.

Figure 4 shows score cluster plot of the eight different human cells with PC1, PC2, and PC3, which were labelled according to their types. Eight human cell groups appeared in cluster trends along the top three PCs axes. PC1 interprets 95.52% variances, PC2 3.17% variances, and PC3 0.66% variances. The cumulative contribution rate of the top three PCs was 99.35%. The 3-dimensional space represented by the top three PCs score indicated 99.35% information from the original spectral data, which covered most of the main information of them.

It could be observed from Figure 4 that there are some inherent component and structure differences among human cell samples even though they actually belong to the same types. The cell samples could not be classified directly using PCA. The separation of the eight different types of samples was not distinct, and especially, some overlapped samples could be examined from the groups of HeLa, HO-8910 and A549. It can be assumed that the biochemical composition of the samples, such as protein, nucleic acid, glycolipid, are similar among the three groups of human cells.

Based on the PCA score plots, geometrical exploration gives the clear clusters trend in the 3D space, instrumental in discriminating types of samples but it cannot be used as a classification tool. Therefore, some discrimination models were used to classify the samples. Supervised pattern recognition approaches refer to some techniques with which a priori knowledge about the category membership of calibration samples are used for distinguishing purposes. The classification model is calibrated on training samples set with different categories [26,27]. The performance of the calibration model is evaluated using the prediction or test set. Two discrimination models were attempted comparatively.

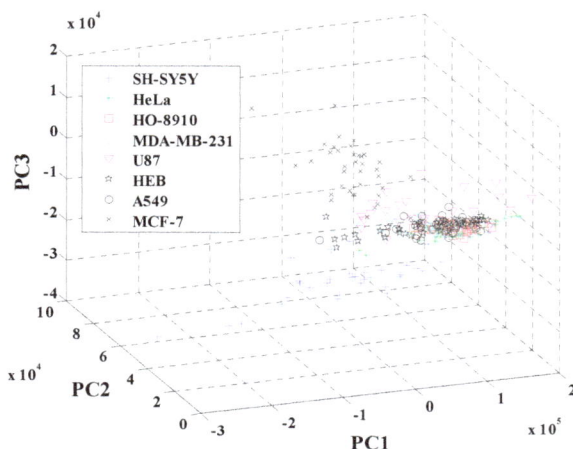

Figure 4. 3-dimensional (3D) space with the top three PCs for the eight different human cells.

3.2. Comparison of Discrimination Models

The Figure 2b gives representative peaks of the A549 cells. Proteins have strong Raman peaks at 1003 cm^{-1} (phenylalanine), 1449 cm^{-1} (CH$_2$ deformation), and 1658 cm^{-1} (Amide I). The main Raman peaks associated nucleic acids correspond to 1176 cm^{-1} (cytosine, guanine), 1339 cm^{-1} (G (DNA/RNA)) and 1580 cm^{-1} (pyrimidine ring of nucleic acids). Lipids have strong Raman peaks at 1304 cm^{-1} (CH$_2$ deformation) and 1449 cm^{-1} (CH$_2$ bending mode in malignant tissue) [28–30]. Raman spectra distinguish the cancer cells coming from different organs, different malignancy degree are all based on the difference of the cellular biochemical components. The basic biochemical components of human cells are consistent, with only a few differences in the chemical composition, so Raman band assignment of the 8 cell lines are similar (Figure 3). Therefore, the identification of 8 types of cells by Raman spectroscopy must be carried out by the means of Chemometrics.

Linear Discriminant Analysis (LDA) and Quadratic Discriminant Analysis (QDA), two classic discriminant methods, were employed in classification of the eight human cells based on Raman data. LDA and QDA are two of the best-known discriminant analysis approaches, which have been successfully used for the appraisement in various fields [30,31]. The boundaries that separate groups or classes of samples are calculated using LDA and QDA. Linear boundaries, where a straight line or hyperplane divides the variable space into regions, and quadratic boundaries, where a quadratic curve divides the variable space into regions, were generated by LDA and QDA, respectively. LDA fits a multivariate normal density to each group, with a pooled estimate of covariance. It does not take into account different variance structures for the two classes. QDA fits multivariate normal densities with covariance estimates stratified by group. It allows for discriminating classes which have significantly different class-specific covariance matrices and forms a separate variance model for each class. QDA classifier focuses on finding a transformation of the input features which is able to optimally distinguish between the different classes in the dataset [32–38].

Figure 5 shows the identification results of LDA and QDA models with 12 PCs. Sample numbers of the eight cell lines for LDA and QDA models with 12 PCs, is shown in Table 2. For the LDA model, the classification rate by cross-validation was 91.21%. A few samples from several groups were wrongly classified in the prediction set; for the QDA model, the classification rate was 100%, which is better than LDA model. The classification results of LDA and QDA models influenced by the number of PCs are presented in Figure 6. As shown in Figure 6, the optimal LDA model was obtained when 15 PCs were used; the optimal QDA model were generated when 12 PCs were employed, and QDA consistently gives a relatively high identification rate.

In order to get a discrimination model for human cell types with good performance, LDA and QDA models were attempted comparatively. Identification results from two models in the calibration set and prediction set, is shown in Table 3. Contrasting to the LDA model, the QDA model obtains a comparatively better performance. It indicated that the quadratic information was helpful to improve the classification performance in the prediction set. Investigated between LDA and QDA models, the LDA adopted hyperplane to classify the samples, while the QDA used higher complexity hypersurface as separator [39]. Because of preferable generalization in its theory, QDA results in a better result than LDA model in prediction set. Quadratic discriminant method is stronger in the level of self-learning and self-adjust than linear discriminant method. Therefore, models based on quadratic discriminant analysis often feature superior performance.

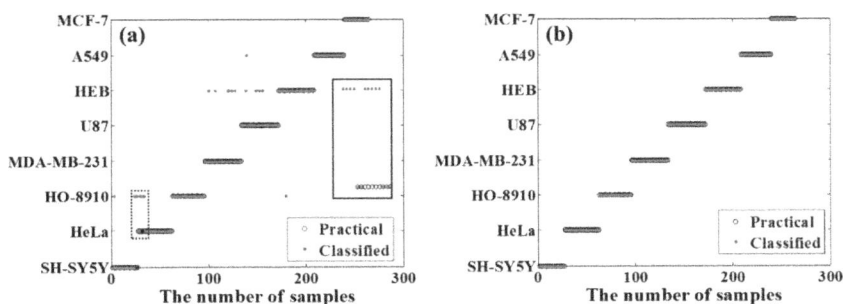

Figure 5. Identification results of the Linear Discriminant Analysis (LDA) model (**a**) and the Quadratic Discriminant Analysis (QDA) model (**b**) with 12 PCs for the eight different human cells. The inset shows the amplified graph to highlight the practical samples and the classified samples clearly.

Table 2. Sample numbers of the eight human cells for Linear Discriminant Analysis (LDA) and Quadratic Discriminant Analysis (QDA) models with 12 PCs.

Samples	Calibration Set	Prediction Set
SH-SY5Y	74	26
HeLa	64	36
HO-8910	67	33
MDA-MB-231	62	38
U87	62	38
HEB	63	37
A549	69	31
MCF-7	75	25
Total	536	264

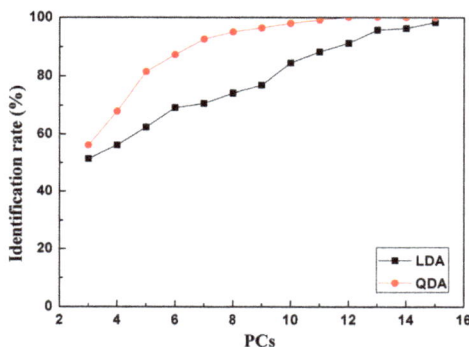

Figure 6. Classification rates of LDA model and QDA model with the number of PCs in the prediction set.

Table 3. Classification rates from LDA model and QDA model with the number of PCs.

| PCs | Classification Rates (%) | | | |
| | LDA | | QDA | |
	Calibration Set	Prediction Set	Calibration Set	Prediction Set
3	53.72	51.39	57.18	56.00
4	57.83	56.15	68.43	67.77
5	64.23	62.42	83.11	81.35
6	68.78	69.21	88.03	87.17
7	71.25	70.68	92.03	92.47
8	75.69	74.20	95.42	94.98
9	78.12	76.85	97.47	96.30
10	83.97	84.47	98.79	97.85
11	90.29	88.26	100.00	99.07
12	93.48	91.21	100.00	100.00
13	96.26	95.75	100.00	100.00
14	98.10	96.35	100.00	100.00
15	100.00	98.35	100.00	100.00

4. Conclusions

Distinguishing of eight different human cells based on Raman spectroscopy was attempted in this work. The PCA method was first attempted to visualize and extract the useful information from multivariate spectral data to examine the qualitative differences among all types of samples. Two discrimination models (LDA and QDA) were developed comparatively in this work. The results indicated that the human cell detection based on Raman spectroscopy was feasible, and the QDA method performed much better in contrast to the LDA method, resulting in the unambiguous identifications of all eight cells. It could be concluded that it is a promising method using Raman spectroscopy technique combined with appropriate discrimination models to distinguish different cancerous human cells. Furthermore, it will be a very interesting topic to study the Raman detection of cancerous cells mixed with the corresponding normal cells (lung, breast, etc.) in our future research.

Acknowledgments: The work was financial supported by National Natural Science Foundation of China (21407145), and National Key Research and Development Program of China (2016YFC0101301).

Author Contributions: Mingjie Tang, Liangping Xia, Dongshan Wei, Shihan Yan, Chunlei Du and Hong-Liang Cui designed and conceived this study. Mingjie Tang and Liangping Xia performed the experiments. Mingjie Tang and Liangping Xia wrote this paper. All authors approved this manuscript.

Conflicts of Interest: The authors declare no conflict of interest.

References

1. Niederhuber, J.E. Global Cancer Control: An Essential Duty. Available online: http://www.cancer.gov/aboutnci/ncicancerbulletin/archive/2009/100609/page4 (accessed on 18 May 2012).

2. Fagerlin, A.; Rovner, D.; Stableford, S.; Jentoft, C.; Wei, J.T.; Holmesrovner, M. Patient education materials about the treatment of early-stage prostate cancer: A critical review. *Ann. Intern. Med.* **2004**, *140*, 721–728. [CrossRef] [PubMed]

3. Winnard, P.T., Jr.; Zhang, C.; Vesuna, F.; Kang, J.W.; Garry, J.; Dasari, R.R.; Barman, I.; Raman, V. Organ-specific isogenic metastatic breast cancer cell lines exhibit distinct Raman spectral signatures and metabolomes. *Oncotarget* **2017**, *8*, 20266–20287. [CrossRef] [PubMed]

4. Das, S.; Batra, S.K. Understanding the Unique Attributes of MUC16 (CA125): Potential Implications in Targeted Therapy. *Cancer Res.* **2015**, *75*, 4669–4674. [CrossRef] [PubMed]

5. Jones, S.; Anagnostou, V.; Lytle, K.; Parpartli, S.; Nesselbush, M.; Riley, D.R.; Shukla, M.; Chesnick, B.; Kadan, M.; Papp, E.; et al. Personalized genomic analyses for cancer mutation discovery and interpretation. *Sci. Transl. Med.* **2015**, *7*, 283ra53. [CrossRef] [PubMed]

6. Ling, J.; Weitman, S.D.; Miller, M.A.; Moore, R.V.; Bovik, A.C. Direct Raman imaging techniques for study of the subcellular distribution of a drug. *Appl. Opt.* **2002**, *41*, 6006–6017. [CrossRef] [PubMed]

7. Notingher, L.; Jell, G.; Notingher, P.L.; Bisson, I.; Tsigkou, O.; Polak, J.M.; Stevens, M.M.; Hench, L.L. Multivariate analysis of Raman spectra for in vitro non-invasive studies of living cells. *J. Mol. Struct.* **2005**, *744*, 179–185. [CrossRef]

8. Owen, C.A.; Notingher, I.; Hill, R.; Stevens, M.; Hench, L.L. Progress in Raman spectroscopy in the fields of tissue engineering, diagnostics and toxicological testing. *J. Mater. Sci. Mater. Med.* **2006**, *17*, 1019–1023. [CrossRef] [PubMed]

9. Huang, J.; Liu, S.; Chen, Z.; Pang, F.; Wang, T. Distinguishing Cancerous Liver Cells Using Surface-Enhanced Raman Spectroscopy. *Technol. Cancer Res. Treat.* **2016**, *15*, 36–43. [CrossRef] [PubMed]

10. Yu, C.X.; Gestl, E.; Eckert, K.; Allara, D.; Irudayaraj, J. Characterization of human breast epithelial cells by confocal Raman micro spectroscopy. *Cancer Detect. Prev.* **2006**, *30*, 515–522. [CrossRef] [PubMed]

11. McEwen, G.D.; Wu, Y.; Tang, M.; Qi, X.; Xiao, Z.; Baker, S.M.; Yu, T.; Gilbertson, T.A.; DeWald, D.B.; Zhou, A. Subcellular spectroscopic markers, topography and nanomechanics of human lung cancer and breast cancer cells examined by combined confocal Raman microspectroscopy and atomic force microscopy. *Analyst* **2013**, *138*, 787–797. [CrossRef] [PubMed]

12. Ball, D.W. Theory of Raman spectroscopy. *Spectroscopy* **2001**, *16*, 32–34.

13. Abramczyk, H.; Surmacki, J.; Kopeć, M.; Olejnik, A.K.; Lubecka-Pietruszewska, K.; Fabianowska-Majewska, K. The role of lipid droplets and adipocytes in cancer. Raman imaging of cell cultures: MCF10A, MCF7, and MDA-MB-231 compared to adipocytes in cancerous human breast tissue. *Analyst* **2015**, *140*, 2224–2235. [CrossRef] [PubMed]

14. Cervo, S.; Mansutti, E.; Mistro, G.D.; Spizzo, R.; Colombatti, A.; Steffan, A.; Sergo, V.; Bonifacio, A. SERS analysis of serum for detection of early and locally advanced breast cancer. *Anal. Bioanal. Chem.* **2015**, *407*, 7503–7509. [CrossRef] [PubMed]

15. Tsikritsis, D.; Richmond, S.; Stewart, P.; Elfick, A.; Downes, A. Label-free identification and characterization of living human primary and secondary tumour cells. *Analyst* **2015**, *140*, 5162–5168. [CrossRef] [PubMed]

16. Kallaway, C.; Almond, L.M.; Barr, H.; Wood, J.; Hutchings, J.; Kendall, C.; Stone, N. Advances in the clinical application of Raman spectroscopy for cancer diagnostics. *Photodiagn. Photodyn. Ther.* **2013**, *10*, 207–219. [CrossRef] [PubMed]

17. Lyng, F.M.; Faoláin, E.Ó.; Conroy, J.; Meade, A.D.; Knief, P.; Duffy, B.; Hunter, M.B.; Byrne, J.M.; Kelehan, P.; Byrne, H.J. Vibrational spectroscopy for cervical cancer pathology, from biochemical analysis to diagnostic tool. *Exp. Mol. Pathol.* **2007**, *82*, 121–129. [CrossRef] [PubMed]

18. Abramczyk, H.; Surmacki, J.; Brożek-Płuska, B.; Morawiec, Z.; Tazbir, M. The hallmarks of breast cancer by Raman spectroscopy. *J. Mol. Struct.* **2009**, *924*, 175–182. [CrossRef]

19. Krishna, C.M.; Sockalingum, G.D.; Kegelaer, G.; Rubin, S.; Kartha, V.B.; Manfait, M. Micro-Raman spectroscopy of mixed cancer cell populations. *Vib. Spectrosc.* **2005**, *38*, 95–100. [CrossRef]

20. Short, K.W.; Carpenter, S.; Frever, J.P.; Mourant, J.R. Raman spectroscopy detects biochemical changes due to proliferation in mammalian cell cultures. *Biophys. J.* **2005**, *88*, 4274–4288. [CrossRef] [PubMed]

21. Notingher, I. Raman Spectroscopy cell-based Biosensors. *Sensors* **2007**, *7*, 1343–1358. [CrossRef]
22. Chen, M.Z.; McReynolds, N.; Campbell, E.C.; Mazilu, M.; Barbosa, J.; Dholakia, K.; Powis, S.J. The Use of Wavelength Modulated Raman Spectroscopy in Label-Free Identification of T Lymphocyte Subsets, Natural Killer Cells and Dendritic Cells. *PLoS ONE* **2015**, *10*, e0125158. [CrossRef] [PubMed]
23. Zhao, J.; Lui, H.; Mclean, D.I.; Zeng, H. Automated autofluorescence background subtraction algorithm for biomedical Raman spectroscopy. *Appl. Spectrosc.* **2007**, *61*, 1225–1232. [CrossRef] [PubMed]
24. O'Farrell, M.; Lewis, E.; Flanagan, C.; Lyons, W.B.; Jackman, N. Combining principal component analysis with an artificial neural network to perform online quality assessment of food as it cooks in a large-scale industrial oven. *Sens. Actuators B-Chem.* **2005**, *107*, 104–112. [CrossRef]
25. Roggo, Y.; Duponchel, L.; Huvenne, J.P. Comparison of supervised pattern recognition methods with McNemar's statistical test—Application to qualitative analysis of sugar beet by near-infrared spectroscopy. *Anal. Chim. Acta* **2003**, *477*, 187–200. [CrossRef]
26. Luo, Y.F.; Guo, Z.F.; Zhu, Z.Y.; Wang, C.P.; Jiang, H.Y.; Han, B.Y. Studies on ANN models of determination of tea polyphenol and amylose in tea by near-infrared spectroscopy. *Spectrosc. Spectr. Anal.* **2005**, *25*, 1230–1233.
27. Liu, C.M.; Fu, S.Y. Effective protocols for kNN search on broadcast multi-dimensional index trees. *Inf. Syst.* **2008**, *33*, 18–35. [CrossRef]
28. Chan, J.W.; Taylor, D.S.; Zwerdling, T.; Lane, S.M.; Ihara, K.; Huser, T. Micro-Raman spectroscopy detects individual neoplastic and normal hematopoietic cells. *Biophys. J.* **2006**, *90*, 648–656. [CrossRef] [PubMed]
29. Shetty, G.; Kendall, C.; Shepherd, N.; Stone, N.; Barr, H. Raman spectroscopy: Elucidation of biochemical changes in carcinogenesis of oesophagus. *Br. J. Cancer* **2006**, *94*, 1460–1464. [CrossRef] [PubMed]
30. Stone, N.; Kendall, C.; Smith, J.; Crow, P.; Barr, H. Raman spectroscopy for identification of epithelial cancers. *Faraday Discuss.* **2004**, *126*, 141–157. [CrossRef] [PubMed]
31. Shao, W.H.; Li, Y.; Diao, S.; Jiang, J.; Dong, R. Rapid classification of Chinese quince (*Chaenomeles speciosa* Nakai) fruit provenance by near-infrared spectroscopy and multivariate calibration. *Anal. Bioanal. Chem.* **2017**, *409*, 115–120. [CrossRef] [PubMed]
32. Siqueira, L.F.S.; Júnior, R.F.A.; Araújo, A.A.D.; Morais, C.L.M.; Lina, K.M.G. LDA vs. QDA for FT-MIR prostate cancer tissue classification. *Chemom. Intell. Lab. Syst.* **2017**, *162*, 123–129. [CrossRef]
33. Dixon, S.J.; Brereton, R.G. Comparison of performance of five common classifiers represented as boundary methods: Euclidean Distance to Centroids, Linear Discriminant Analysis, Quadratic Discriminant Analysis, Learning Vector Quantization and Support Vector Machines, as dependent on data structure. *Chemom. Intell. Lab. Syst.* **2009**, *95*, 1–17.
34. Ali, S.; Veltri, R.; Epstein, J.I.; Christudass, C.; Madabhushi, A. Selective invocation of shape priors for deformable segmentation and morphologic classification of prostate cancer tissue microarrays. *Comput. Med. Imaging Graph.* **2015**, *41*, 3–13. [CrossRef] [PubMed]
35. Viswanath, S.E.; Bloch, N.B.; Chappelow, J.C.; Toth, R.; Rofsky, N.M.; Genega, E.M.; Lenkinski, R.E.; Madabhushi, A. Central gland and peripheral zone prostate tumors have significantly different quantitative imaging signatures on 3 tesla endorectal, in vivo T2-weighted MR imagery. *J. Magn. Reson. Imaging* **2012**, *36*, 213–224. [CrossRef] [PubMed]
36. Wu, W.; Mallet, Y.; Walczak, B.; Penninckx, W.; Massart, D.L.; Heuerding, S.; Erni, F. Comparison of regularized discriminant analysis, linear discriminant analysis and quadratic discriminant analysis, applied to NIR data. *Anal. Chim. Acta* **1996**, *329*, 257–265. [CrossRef]
37. Pamukcu, E.; Bozdogan, H.; Caljk, S. A Novel Hybrid Dimension Reduction Technique for Undersized High Dimensional Gene Expression Data Sets Using Information Complexity Criterion for Cancer Classification. *Comput. Math. Methods Med.* **2015**, *2015*, 370640. [CrossRef] [PubMed]
38. Zhang, H.Y.; Wang, H.; Dai, Z.; Chen, M.S.; Yuan, Z. Improving accuracy for cancer classification with a new algorithm for genes selection. *BMC Bioinform.* **2012**, *13*, 298. [CrossRef] [PubMed]
39. Ba, Y.T.; Zhang, W.; Wang, Q.; Zhou, R.; Ren, C. Crash prediction with behavioral and physiological features for advanced vehicle collision avoidance system. *Transp. Res. Part C-Emerg. Technol.* **2017**, *74*, 22–33. [CrossRef]

applied sciences

MDPI

Article

A Geometric Dictionary Learning Based Approach for Fluorescence Spectroscopy Image Fusion

Zhiqin Zhu [1,2], Guanqiu Qi [1,3,*], Yi Chai [2] and Penghua Li [1]

[1] College of Automation, Chongqing University of Posts and Telecommunications, Chongqing 400065, China; zhiqinzhu@outlook.com (Z.Z.); lipenghua88@163.com (P.L.)

[2] State Key Laboratory of Power Transmission Equipment and System Security and New Technology, Chongqing University, Chongqing 400044, China; chaiyi@outlook.com

[3] School of Computing, Informatics, and Decision Systems Engineering, Arizona State University, Tempe, AZ 85287, USA

[*] Correspondence: guanqiuq@asu.edu; Tel:+1-323-742-0133

Academic Editor: Johannes Kiefer
Received: 18 December 2016 ; Accepted: 28 January 2017 ; Published: 9 February 2017

Abstract: In recent years, sparse representation approaches have been integrated into multi-focus image fusion methods. The fused images of sparse-representation-based image fusion methods show great performance. Constructing an informative dictionary is a key step for sparsity-based image fusion method. In order to ensure sufficient number of useful bases for sparse representation in the process of informative dictionary construction, image patches from all source images are classified into different groups based on geometric similarities. The key information of each image-patch group is extracted by principle component analysis (PCA) to build dictionary. According to the constructed dictionary, image patches are converted to sparse coefficients by simultaneous orthogonal matching pursuit (SOMP) algorithm for representing the source multi-focus images. At last the sparse coefficients are fused by Max-L1 fusion rule and inverted to fused image. Due to the limitation of microscope, the fluorescence image cannot be fully focused. The proposed multi-focus image fusion solution is applied to fluorescence imaging area for generating all-in-focus images. The comparison experimentation results confirm the feasibility and effectiveness of the proposed multi-focus image fusion solution.

Keywords: Image Fusion; Sparse Representation; Dictionary Construction; Geometric Classification; Multi-focus; Fluorescence Imaging

1. Introduction

Following the development of cloud computing, cloud environment provides more and more strong computation capacity to process various images [1–3]. Due to the limited depth-of-focus of optical lenses in imaging sensors, one-time focus cannot guarantee to obtain all focused image in the whole scene [4]. In another word, it is difficult to obtain an image that contains all relevant objects in focus. In microscope, only the objects at a certain distance away from the lens can be captured in focus and sharply, and other objects are out of focus and blurred. A fuzzy multi-sensor data fusion Kalman filter model was proposed by Rodger to reduce failure risk in an integrated vehicle health maintenance system (IVHMS) [5]. In fluorescence spectroscopy, one scene at least contains several objects. For ensuring the accuracy and efficiency of fluorescence image processing, it usually captures multiple images of the same scene to guarantee all objects are in focus. However, it costs too much on viewing and analyzing a series of images separately. Multi-focus image fusion is an effective technique to resolve this problem by combining complementary information from multiple images of the same scene into a sharp and accurate one [6,7]. The integrated image can precisely describe the corresponding scene, which is beneficial for further analysis and understanding.

As one of the most widely recognized image fusion technologies, a large number of multi-focus image fusion methods have been proposed. According to fusion domain, these methods could be categorized into two main categories: spatial-domain-based methods and transform-domain-based methods [8]. Spatial-domain-based methods directly choose clear pixels, blocks, or regions from source images to compose a fused image [9,10]. Some simple methods, such as averaging or max pixel schemes, are performed on single pixel to generate a fused image. However, these methods may reduce the contrast and edge intensity of fused result. In order to improve the performance of fused image, some advanced schemes, such as block-based and region-based algorithms, were proposed. Li et al. proposed a scheme by dividing images into blocks and chose the focused one by comparing spatial frequencies (SF) first; then the fused results were produced by consistency verification [11]. Although block-based methods improve the contrast and sharpness of integrated image, they may cause block effect in integrated image [12].

Different from spatial-domain fusion methods, transform-domain methods transform source images into a few corresponding coefficients and transform bases first [13]. Then the coefficients are fused and inverted to an integrated image. Multi-scale transform (MST) and wavelet based algorithms are the most conventional transform approaches applied to transform-domain-based image fusion [14,15]. Wavelet transform [16,17], shearlet [18,19], curvelet [20], dual tree complex wavelet transform [21,22], nonsubsampled controulet transform (NSCT) [23] are usually used in MST-based methods. MST decomposition methods have attracted great attention in image processing field and achieved great performance in image fusion fields. However, MST-based methods are difficult to select an optimal transform basis without priori knowledge [24,25]. As each MST method has its own limitations, one MST method is difficult to fit all kinds of images [14].

In recent years, sparse-representation-based methods, as the subsets of MST fusion methods, are applied to image fusion. Other than MST methods, sparse-representation-based methods usually use trained bases, which can adaptively change according to input images without priori knowledge [26,27]. Due to adaptive learning feature, sparse representation is widely applied to image de-noising [28,29], image deblurring [30,31], image inpainting [32], image fusion [33], and super-resolution [34].

Yang and Li [35] took the first step for applying sparse representation theory to image fusion field. They proposed a multi-focus image fusion method with a fixed-dictionary. Li and Zhang [36] applied morphologically filtering sparse feature to matrix decomposition method for improving the accuracy of sparse representation in image fusion. Wang and Liu [37] proposed an approximate K-SVD based sparse representation method for image fusion and exposure fusion to reduce computation costs. Kim and Han [38] proposed a joint clustering dictionary learning method for image fusion. They used the steering kernel regression to strength the local geometric features of source images first. Then K-means clustering method was applied to image pixels clustering based on the local image features. The proposed method by Kim and Han can significantly group the pixels of input source images into a few classes. The principle components of local features are extracted from each group to construct a dictionary, which can make a compact and informative dictionary. This method reduces the sparse coding time by using the constructed compact dictionary. However, the performance of Kim and Han's method depends on the presetting cluster number, that is difficult to confirm before clustering. The number of local geometric-feature types is usually used as the presetting cluster number, but the actual clustering results do not exactly follow the geometric features every time. So the clustering results cannot reach the expectations all the time. To optimize the weakness of existing clustering methods, a geometric information based image block classification method is proposed in this paper.

Geometric information, as one type of the most important image information, including edges, contours, and textures of image, and so on, can remarkably influence the quality of image perception [39]. The information can be used in patch classification as a supervised dictionary prior to improve the performance of trained dictionary [40,41]. In this paper, a geometric classification based dictionary learning method is proposed for sparse-representation based image fusion. Instead of grouping the pixels of images, the proposed geometric classification method groups image blocks directly by the geometric similarity of each image block. Since sparse-representation based fusion method uses image blocks for sparse coding and coefficients fusion, extracting underlying geometric information from image-block groups is an efficient way to construct a dictionary. Moreover, the geometric classification can group image blocks based on edge, sharp line information for dictionary learning, which can improve the accuracy of sparse representation. The proposed dictionary learning approach consists of three steps:

- In the first step, input source images are split into small blocks. According to the similarity of geometric patterns, these blocks are classified into smooth patches, stochastic patches, and dominant orientation patches.
- In the second step, it does principal component analysis (PCA) on each type of patches to extract crucial bases. The extracted PCA bases are used to construct the sub-dictionary of each image patch group.
- In the last step, the obtained sub-dictionaries are merged into a complete dictionary for sparse-representation-based fusion approach and the integrated sparse coefficients are inverted to fused image.

This paper has two main contributions.

1. A geometric image patches classification method is proposed for dictionary learning. The proposed geometry classification method can accurately split source image patches into different image patch groups for dictionary learning. Dictionary bases extracted from each image patch group have good performance, when they are used to describe the geometric features of source images.
2. A PCA-based geometry dictionary construction method is proposed. The trained dictionary with PCA bases is informative and compact for sparse representation. The informative feature of trained dictionary ensures that different geometric features of source images can be accurately described. The compact feature of trained dictionary can speed up the sparse coding process.

The rest sections of this paper are structured as follows: Section 2 proposes the geometric dictionary learning method and multi-focus image fusion framework; Section 3 compares and analyzes experimentation results; and Section 4 concludes this paper.

2. Geometry Dictionary Construction and Fusion Scheme

2.1. Dictionary Learning Analysis

During the sparse representation of the dictionary construction process, the key step is to construct an over-complete dictionary that covers the important information of input image. Since geometry information of image plays an important role in describing an image, an informative dictionary should take the geometric information into consideration. Smooth and non-smooth information as two important types of geometry information can be used to classify source images.

A multiple geometry information classification approach was proposed for single image super-resolution reconstruction (SISR) [39]. According to the geometric features, a large number of high-resolution (HR) image patches were randomly extracted and clustered to generate corresponding geometric dictionaries for sparse representations of local patches in low-resolution (LR) images. The geometric features were classified into three types, smooth patch, dominant orientation patch, and stochastic patch. According to the estimated dominant orientation, the dominant orientation patches could be divided into different directional patches. Rather than adaptively selecting one dictionary, the recovered patches were weighted in the learned geometric dictionaries to characterize the local image patterns, by a subsequent patch aggregation to estimate the whole image. To reduce the redundancy in image recovery, the self-similarity constraints were added on HR image patch aggregation. Both LR and HR residual images were estimated from the recovered image and compensated for the subtle details of reconstructed image.

In SISR, the source images are classified as smooth, dominant orientation, and stochastic patches. Dominant orientation and stochastic patches are non-smooth patches. Since the detection of orientation in finite dimensional Euclidean spaces corresponded to fitting an axis or a plane by Fourier transformation of an n-dimensional structure, Bigün verified dominant orientation and stochastic patches can be separated by orientation estimation in the spatial domain based on the error of the fit [42].

In dictionary learning, the redundancy of trained dictionary is usually not considered. A redundant dictionary not only has high computational complexity, but cannot also obtain the best effect in image representation. Different methods are proposed to reduce the redundancy in learning process to construct the compact dictionary. Elad [43] estimated the maximum a posteriori probability (MAP) by a compact dictionary. An effective image interpolation method by nonlocal autoregressive modeling (NARM) was developed and embedded in SRM by Dong [44] to enhance the effectiveness of SRM in image interpolation by reducing the coherence between the sampling matrix and the sparse representation dictionary and minimizing the nonlocal redundancy. To improve the performance of sparse representation-based image restoration, a non-locally centralized sparse representation (NCSR) model proposed by Dong [45] suppressed the sparse coding noise in image restoration.

The sparse representation-based approach linearly combines a few atoms extracted from an over-complete dictionary as an image patch. The promising results have been shown in various image restoration applications [38,45]. Based on the classification of image patches, this paper proposed sparse representation-based approach that uses PCA algorithm to construct more informative and compact dictionary [38]. The proposed solution cannot be extended to other sparse-representation applications. Most sparse representation methods are based on a large number of sample images. The corresponding learned dictionaries are expected to be applied to different scenic source images. The proposed solution uses sparse feature to do image sparse representation and fusion. Therefore, the learned dictionary of proposed solution does not need to be extended to different source images, and only specializes for experimented source images. It uses PCA classification to reduce the redundancy in dictionary and enhance the performance.

There are many popular dictionary learning methods like KSVD. However, this paper cannot use other dictionary learning methods to substitute PCA to do dictionary learning. It compares PCA with KSVD to explain the benefits of using PCA in dictionary learning. For sparse representation-based image reconstruction and fusion techniques, it is difficult to select a good over-complete dictionary. To obtain an adaptive dictionary, Aharon [46] developed KSVD that learned an over-complete dictionary by using a set of training images or updates a dictionary adaptively by applying SVD operations. The image patches used by KSVD in dictionary learning were extracted either from globally trained dictionary (natural images) or adaptively trained dictionary (input images). Although KSVD shows better performance in dictionary learning than other methods, it takes high costs to do dictionary learning with a large number of training images iteratively. The high computational complexity constrains the learned dictionary size of KSVD in practical usage. Kim [38] firstly applied clustering-based dictionary learning solution to image fusion. It clustered patches from different source images together based on local structural similarities. Only a few principal components of each cluster were analyzed and used to construct corresponding sub-dictionary. All learned sub-dictionaries were combined to form a compact and informative dictionary that can describe local structures of all source images effectively. The redundancies of dictionary (i.e., reducing the size of dictionary) were eliminated to reduce the computation loads of the proposed PCA-based dictionary learning solution. In various comparison experimentations, the proposed PCA-based dictionary learning solution (JPDL) had better performance than other traditional multi-scale transform-based and sparse representation-based methods, including KSVD. Comparing with KSVD, JPDL as a PCA-based method not only obtained a more compact dictionary, but also took less computational costs in dictionary learning process.

2.2. Geometry Dictionary Construction

According to geometric patterns used in SISR [39], this paper classifies the image into three patch types: smooth patches, stochastic patches, and dominant orientation patches. The smooth patches, stochastic patches and dominant orientation patches mainly describe the structure information, texture information, and edge information respectively. Based on a priori knowledge, image patches can be classified into three groups for sub-dictionaries learning. Additionally, to obtain more compact sub-dictionary for each image patch group, this paper uses PCA method to extract the important information from each group. PCA bases of each group are used as the bases of corresponding sub-dictionary. The obtained dictionary can be more compact and informative by combining the PCA-based sub-dictionaries [38].

This paper learns three different sub-dictionaries rather other one dictionary. Three different sub-dictionaries contain more detailed information and structure information, and less redundant information. It means that although the obtained sub-dictionaries are compact, they can still contain more useful information. Three types of geometric patches extract different image information from source images respectively.

- **Smooth** patches describe the structure information of source images, such as the background.
- **Dominant orientation** patches describe the edge information to provide the direction information in source images.
- **Stochastic** patches show the texture information to make up the missing detailed information that is not represented in dominant orientation patches.

Three learned dictionaries specialize in smooth, dominant orientation, and stochastic patches respectively. The structure information, edge information, and texture information can be accurately and completely shown in corresponding sub-dictionary. Comparing with one learned dictionary, three learned sub-dictionaries can not only supply different information of source images respectively, but also make up the deficiencies mutually to enhance the quality of merged dictionary. So each sub-dictionary is important to compose a compact and informative dictionary in the proposed approach.

The proposed geometric approach is shown in Figure 1 that has two main steps. In the first step, the input source images I_i to I_k are split to several small image blocks $p_{in}, i \in (1, 2...k), n \in (1, 2...w)$, where i is the source image number, and n is the patch number. The total block number of each input image is w. Then according to the similarity of geometric patterns, the obtained image blocks are classified into three groups, smooth patch group, stochastic patch group, and dominant orientation patch group. In the second step, it does PCA analysis on each sub-class for extracting corresponding PCA bases as sub-dictionary. Then these obtained sub-dictionaries are composed to construct a complete dictionary for image sparse representation.

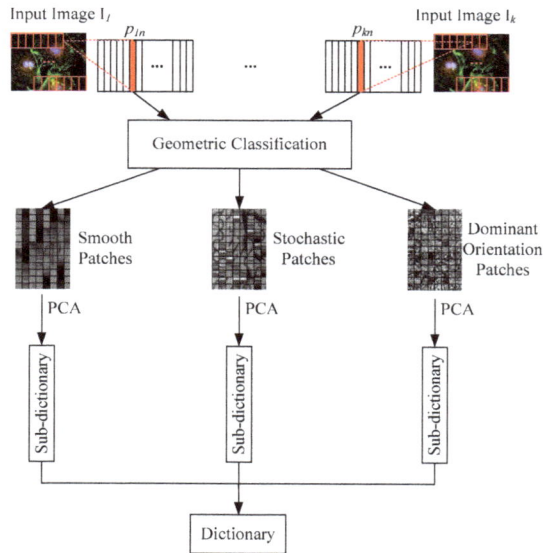

Figure 1. Geometric Dictionary based Image Fusion Framework.

2.3. Geometric-Structure-Based Patches Classification

Image blocks of a source image can be classified by various image features in describing underlying relationships. According to geometry descriptions, the multi-focus source images can be divided to smooth, stochastic, and dominant orientation patches. The edges of focused area are usually sharp that contain dominant orientation patches. The out-of-focused areas are smooth that contain a large number of smooth blocks. Besides that, there are also lots of stochastic blocks in source images. Grouping image blocks into different classes for dictionary learning is an efficient way to improve the accuracy of image description.

To obtain different geometry sub-classes, this paper uses a geometric-structure-based approach to partition image into several sub-classes first. Then based on the classified image blocks, it gets the corresponding sub-dictionaries.

Before doing geometry analysis, it needs to divide the input image into $\sqrt{w} \times \sqrt{w}$ small image blocks $P_I = (p_1, p_2, ..., p_n)$ first. Each image patch $p_i, i \in (1, 2, ...n)$ is converted into $1 \times w$ image vectors $v_i, i \in (1, 2, ...n)$. After obtaining vectors, the variance C_i of pixels in each image vector can be calculated. After obtaining variance, it needs to choose the threshold δ to evaluate whether image block is smooth. If $C_i < \delta$, image block p_i is smooth, otherwise image block p_i is not smooth [39]. Based on the threshold δ, the classified smooth and non-smooth patches are shown in Figure 2a,b respectively.

According to the classified smooth and non-smooth patches shown in Figure 2a,b, it is clear to find that the smooth patches have similar structure information of input images. Oppositely, non-smooth patches are different and cannot be directly classified into one class.

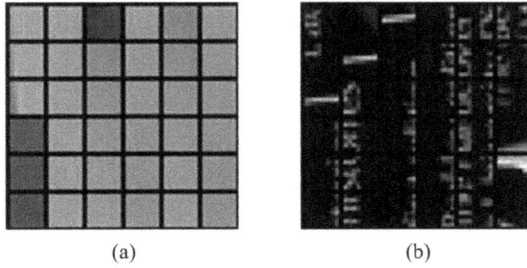

(a)   (b)

Figure 2. Smooth Image Patches and Non-smooth Image Patches, (a) shows smooth image patches, (b) shows non-smooth image patches.

Non-smooth patches may also be classified. According to geometric patterns, non-smooth patches could be divided into stochastic and dominant orientation patches. The separation method of stochastic and dominant orientation image patches consists of two steps. In the first step, the gradient of each pixel is calculated. In every image vector v_i, $i \in (1, 2, ... n)$, the gradient of each pixel k_{ij}, $j \in (1, 2, ..., w)$, $i \in (1, 2, ... n)$ is composed by its x and y coordinate gradient $g_{ij}(x)$ and $g_{ij}(y)$. The gradient value of each pixel k_{ij} in image patch v_i is $g_{ij} = (g_{ij}(x), g_{ij}(y))$. The $(g_{ij}(x), g_{ij}(y))$ can be calculated by $g_{ij}(x) = \partial k_{ij}(x, y)/\partial x$, $g_{ij}(y) = \partial k_{ij}(x, y)/\partial y$. For each image vector v_i, the gradient G_i is $G_i = (g_{i1}, g_{i2}, ..., g_{iw})^T$, where $G_i \in \mathbb{R}^{w \times 2}$. In the second step, the gradient value of each image patch is decomposed by Equation (1):

$$G_i = \begin{bmatrix} G_{i1} \\ G_{i2} \\ ... \\ G_{iw} \end{bmatrix} = U_i S_i V_i^T , \tag{1}$$

where $U_i S_i V_i^T$ is the singular value decomposition of G_i. S_i is a diagonal 2×2 matrix for representing energy in dominant directions [47]. When S_i is obtained, the dominant measure R can be calculated. The calculation method of R is shown in Equation (2):

$$R = \frac{S_{1,1} - S_{2,2}}{S_{1,1} + S_{2,2}} , \tag{2}$$

The smaller the R is, the more stochastic the corresponding image patch is [48]. In this case, a threshold R^* should be calculated for distinguishing stochastic and dominant orientation patches. To find the threshold R^*, a probability density function (PDF) of R is calculated. According to [42], the PDF of R can be calculated by Equation (3).

$$P(R) = 4(w - 1)R \frac{(1 - R^2)^{w-2}}{(1 + R^2)^w} , \tag{3}$$

The PDF of dominant measure R of patches with different sizes is shown in Figure 3.

A PDF significance test is implemented to distinguish stochastic and dominant orientation patch by the threshold R^* [42]. If R is less than R^*, the image patch can be considered as stochastic patch. The stochastic and dominant orientation patches separated by the proposed method are shown in Figure 4. The Figure 4a shows the stochastic image patches, which contain some texture and detailed information. Dominant orientation image patches are shown in Figure 4b, which mainly contain the edge information.

As shown in Figure 3, when R increases, $P(R)$ converges to zero. It chooses the value of R as the threshold R^*, when $P(R)$ reaches zero for the first time.

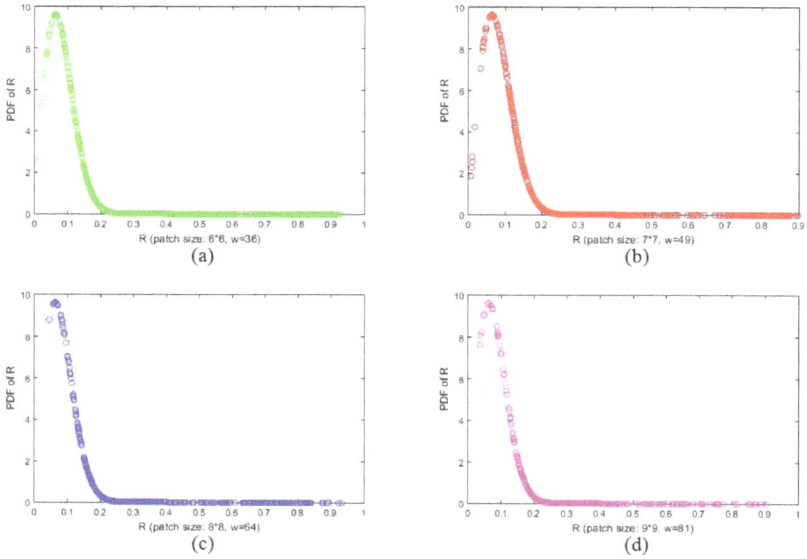

Figure 3. PDFs of Dominant Measure R, (a) shows PDF of R in 6*6 patch size, (b) shows PDF of R in 7*7 patch size, (c) shows PDF of R in 8*8 patch size, (d) shows PDF of R in 9*9 patch size.

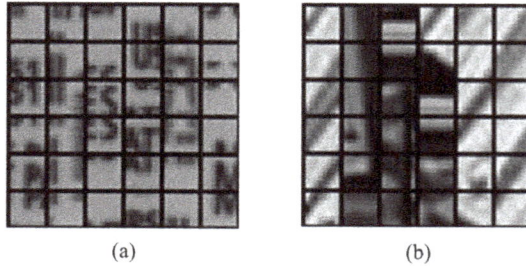

Figure 4. Stochastic Image Patches and Dominant Orientation Image Patches, (a) shows stochastic image patches, (b) shows dominant orientation image patches.

2.4. PCA-Based Dictionary Construction

When the geometric classification finished, image patches with similar geometric structure are classified into a few groups. In this work, the compact and informative dictionary is trained by combining the principal components of each geometric group. Since patches in the same geometric group can be well-approximated by a small number of PCA bases, top m most informative principal components are chosen to form a sub-dictionary [49] as Equation (4).

$$B_x = [b_1, b_2..., b_j], \quad s.t. \quad p = \arg\max_p \left\{ \sum_{j=p+1}^{q} L_j > \delta \right\}, \tag{4}$$

where B_x denotes the sub-dictionary of the xth cluster, and q is the total number of atoms in each cluster. Each B_x consists of p eigenvector atoms. L_j is the eigenvalue of the corresponding jth eigenvector d_j. The eigenvalues are sorted in descending order (i.e., $L_1 > L_2 > ... > L_n > 0$). δ is a parameter to control the amount of approximation with rank p. Usually, δ is set to a proportional to the dimension of input image to avoid over-fitting [49]. After the sub-dictionaries are obtained, a dictionary D is constructed by combining sub-dictionaries together as Equation (5).

$$D = [B_1, B_2 ..., B_n] \ , \tag{5}$$

where n is the total number of clusters.

2.5. Fusion Scheme

The fusion scheme of proposed method is shown in Figure 5, which consists of two steps.

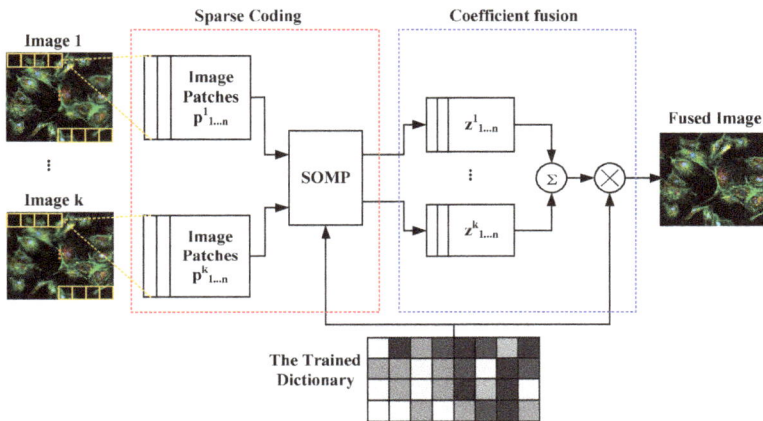

Figure 5. Proposed Image Fusion Scheme.

In the first step, each input image I_i is split into n image patches with the size of $\sqrt{w} \times \sqrt{w}$. These image patches are resized to $w * 1$ vectors $p_1^i, p_2^i, ..., p_n^i$. The resized vectors are sparse coded with trained dictionary to sparse coefficients $z_1^i, z_2^i, ..., z_n^i$. In the second step, the sparse coefficients are fused by 'Max-L1' fusion rule [50–52]. Then the fused coefficients are inverted to fused image by the trained dictionary.

3. Experiments and Analyses

In comparison experiments, three pairs of color fluorescence images are used to test the proposed multi-focused image fusion approach. All three multi-focus image pairs are modified to the size of 256×256 for comparison purpose. To show the efficiency of proposed method, the state of art dictionary learning based sparse-representation fusion schemes KSVD [50] and JPDL [38], which were proposed by Li in 2012 and Kim in 2016 respectively, are used in comparison experiments. The comparison experiments are evaluated by both subjective and objective assessments. Four popular image fusion quality metrics are used in this paper for the quantitative evaluation. The patch size of all sparse-representation-based methods including the proposed method are set to 8×8. To avoid blocking artifacts, all experiments use sliding window scheme [38,50]. The overlapped region of sliding window scheme is set to 4-pixel in each vertical and horizontal direction of all experiments. All experiments were implemented using Matlab, version 2014a; MathWorks: Natick, MA, 2014, and Visual Studio, version Community 2013; Microsoft: Redmond, WA, 2013, mixed programming on an Intel(R) Core(TM), version i7-4720HQ; Intel: Santa Clara, CA, 2015, CPU @ 2.60 GHz Laptop with 12.00 GB RAM.

3.1. Objective Evaluation Methods

Five mainstream objective evaluation metrics are implemented for the quantitative evaluation. These metrics include edge retention ($Q^{AB/F}$) [53,54], mutual information (MI) [55], visual information fidelity (VIF) [56], Yang proposed fusion metric (Q_Y) [57,58], and Chen-Blum metric (Q_{CB}) [58,59]. For the fused image, the sizes of $Q^{AB/F}$, MI, VIF, Q_Y, and Q_{CB} become bigger, the corresponding fusion results are better.

3.1.1. Mutual Information

MI for images can be formalized as Equation (6).

$$MI = \sum_{i=1}^{L} \sum_{j=1}^{L} h_{A,F}(i,j) log_2 \frac{h_{A,F}(i,j)}{h_A(i)h_{F(j)}} \quad , \tag{6}$$

where L is the number of gray-level, $h_{R,F}(i,j)$ is the gray histogram of image A and F. $h_A(i)$ and $h_F(j)$ are edge histogram of image A and F. MI of fused image can be calculated by Equation (7).

$$MI(A,B,F) = MI(A,F) + MI(B,F) \quad , \tag{7}$$

where $MI(A,F)$ represents the MI value of input image A and fused image F; $MI(B,F)$ represents the MI value of input image B and fused image F.

3.1.2. $Q^{AB/F}$

$Q^{AB/F}$ metric is a gradient-based quality index to measure how well the edge information of source images conducted to the fused image. It is calculated by:

$$Q^{AB/F} = \frac{\sum_{i,j} (Q^{AF}(i,j)w^A(i,j) + Q^{BF}(i,j)w^B(i,j))}{\sum_{i,j} (w^A(i,j) + w^B(i,j))} \quad , \tag{8}$$

where $Q^{AF} = Q_g^{AF} Q_0^{AF}$, Q_g^{AF} and Q_0^{AF} are the edge strength and orientation preservation values at location (i, j). Q^{BF} can be computed similarly to Q^{AF}. $w_A(i,j)$ and $w_B(i,j)$ are the weights of Q^{AF} and Q^{BF} respectively.

3.1.3. Visual Information Fidelity

VIF is the novel full reference image quality metric. VIF quantifies the mutual information between the reference and test images based on Natural Scene Statistics (NSS) theory and Human Visual System (HVS) model. It can be expressed as the ratio between the distorted test image information and the reference image information, the calculation equation of VIF is shown in Equation (9).

$$VIF = \frac{\sum_{i \in subbands} I(\overrightarrow{C^{N,i}}; \overrightarrow{F^{N,i}})}{\sum_{i \in subbands} I(\overrightarrow{C^{N,i}}; \overrightarrow{E^{N,i}})} \quad , \tag{9}$$

where $I(\overrightarrow{C^{N,i}}; \overrightarrow{F^{N,i}})$ and $I(\overrightarrow{C^{N,i}}; \overrightarrow{E^{N,i}})$ represent the mutual information, which are extracted from a particular subband in the reference and the test images respectively. $\overrightarrow{C^N}$ denotes N elements from a random field, $\overrightarrow{E^N}$ and $\overrightarrow{F^N}$ are visual signals at the output of HVS model from the reference and the test images respectively.

To evaluate the VIF of fused image, an average of VIF values of each input image and the integrated image is proposed [56]. The evaluation function of VIF for image fusion is shown in Equation (10).

$$VIF(A, B, F) = \frac{VIF(A, F) + VIF(B, F)}{2} \quad , \tag{10}$$

where $VIF(A, F)$ is the VIF value between input image A and fused image F; $VIF(B, F)$ is the VIF value between input image B and fused image F.

3.1.4. Q_Y

Yang et al. proposed structural similarity-Based way for fusion assessment [57]. The Yang's method is shown in Equation (11).

$$Q_Y = \begin{cases} \lambda(\omega)SSIM(A, F \,|\omega) + (1 - \lambda(\omega))(SSIM(B, F \,|\omega), \\ SSIM(A, B \,|\omega) \geq 0.75, \\ max\left\{SSIM(A, F \,|\omega), SSIM(A, B \,|\omega)\right\}, \\ SSIM(A, B \,|\omega) < 0.75 \end{cases} \quad , \tag{11}$$

where $\lambda(\omega)$ is the local weight, $SSIM(A, B)$ is a structural similarity index measure for images A and B. The detail of $\lambda(\omega)$ and $SSIM(A, B)$ can be found in [57,58].

3.1.5. Q_{CB}

Chen-Blum metric is human perception inspired fusion metrics. Chen-Blum metric consists of 5 steps:

In the first step, filtering image $I(i, j)$ in frequency domain. $I(i, j)$ is transformed to frequency domain and get $I(m, n)$. Filtering $I(m, n)$ by contrast sensitive function(CSV) [59] filter $S(r)$, where $r = \sqrt{m^2 + n^2}$. In this image fusion metric $S(r)$ is in polar form. $\tilde{I}(m, n)$ can be got by $\tilde{I}(m, n) = I(m, n) * S(r)$.

In the second step, local contrast is computed. For Q_{CB} metric, Peli's contrast is used and it can be defined as:

$$C(i, j) = \frac{\phi_k(i, j) * I(i, j)}{\phi_{k+1}(i, j) * I(i, j)} - 1 \quad . \tag{12}$$

A common choice for $\phi_k(i, j)$ would be

$$\phi_k(i, j) = \frac{1}{\sqrt{2\pi}\sigma_k} e^{\frac{i^2 + j^2}{2\sigma_k^2}} \quad . \tag{13}$$

with a standard deviation $\sigma_k = 2$.

In the third step, The masked contrast map for input image $I_A(i, j)$ is calculated as:

$$C'_A = \frac{t(C_A)^p}{h(C_A)^p + Z} \quad . \tag{14}$$

Here, t, h, p, q and Z are real scalar parameters that determine the shape of the nonlinearity of the masking function [59,60].

In the fourth step, the saliency map of $I_A(i, j)$ can be calculated by Equation (15),

$$\lambda_A(i, j) = \frac{{C'_A}^2(i, j)}{{C'_A}^2(i, j) + {C'_B}^2(i, j)} \quad . \tag{15}$$

The information preservation value is computed as Equation (16),

$$Q_{AF}(i,j) = \begin{cases} \frac{C'_A(i,j)}{C'_F(i,j)}, if C'_A(i,j) < C'_F(i,j), \\ \frac{C'_F(i,j)}{C'_A(i,j)}, otherwise. \end{cases} \tag{16}$$

In the fifth step, the Global quality map can be calculated:

$$Q_{GQM}(i,j) = \lambda_A(i,j)Q_{AF}(i,j) + \lambda_B(i,j)Q_{BF}(i,j) \ . \tag{17}$$

Then the value of Q_{CB} can be got by average the global quality map:

$$Q_{CB} = \overline{\sum_{i,j} Q_{GQM}(i,j)} \ . \tag{18}$$

3.2. Image Quality Comparison

To show the efficiency of proposed method, the quality comparison of fused images is demonstrated. It compares the quality of fused image based on visual effect, the accuracy of focused region detection, and the objective evaluations.

3.2.1. Comparison Experiment 1

The source fluorescence image is obtained from public website [61]. Figure 6a,b are the source multi-focus fluorescence images. To show the details of fused image, two image blocks are highlighted and magnified, which are squared by red and blue square respectively. The image block in red square is out of focus in Figure 6a, and the image block in blue square, is out of focus in Figure 6b. The corresponding image block in blue and red square are totally focused in Figure 6a,b respectively. Figure 6c–e show the fused images of KSVD, JPDL, and proposed method, respectively.

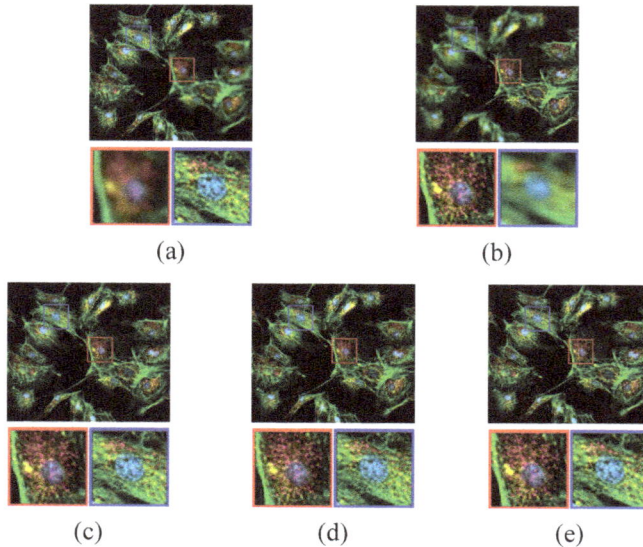

(a)   (b)

(c)   (d)   (e)

Figure 6. Fusion Results of Multi-focus Fluorescence Image - 1, (**a**,**b**) are source images, (**c**–**e**) are fused image of KSVD, JPDL, and proposed method respectively.

The difference and performance of fused images by three different methods are difficult to figure out by eyes. In order to evaluate of fusion performances objectively, $Q^{AB/F}$, MI, VIF, Q_Y, and Q_{CB} are also used as image fusion quality measures. The fusion results of multi-focus fluorescence images using three different methods are shown in Table 1.

Table 1. Fusion Performance Comparison - 1 of Multi-focus Fluorescence Image Pairs.

	$Q^{AB/F}$	**MI**	**VIF**	Q_Y	Q_{CB}
KSVD	0.4966	2.4247	0.5778	0.5960	0.5287
JPDL	0.5815	2.9258	0.6972	0.6944	0.7243
Proposed Solution	**0.6226***	**3.4773**	**0.7428**	**0.7386**	**0.7974**

* The highest result in each column is marked in bold-face.

The best results of each evaluation metric are highlighted by bold-faces in Table 1. According to Table 1, it can figure out that the proposed method has the best performance in all five types of evaluation metrics. Particularly, for the objective evaluation metric $Q^{AB/F}$, the proposed method obtains higher result than other two comparison image fusion methods. Since $Q^{AB/F}$ is a gradient-based quality metric to measure how well the edge information of source images is conducted to the fused image, it means the proposed method can get better fused image with edge information.

3.2.2. Comparison Experiment 2 and 3

Similarly, the source fluorescence images shown in Figures 7 and 8a,b are obtained from public websites [62,63] respectively. In a set of source images, two images (a) and (b) focus on different items. The source images are fused by KSVD, JPDL, and proposed method to get a totally focused image, and the corresponding fusion results are shown in Figures 7 and 8c–e respectively.

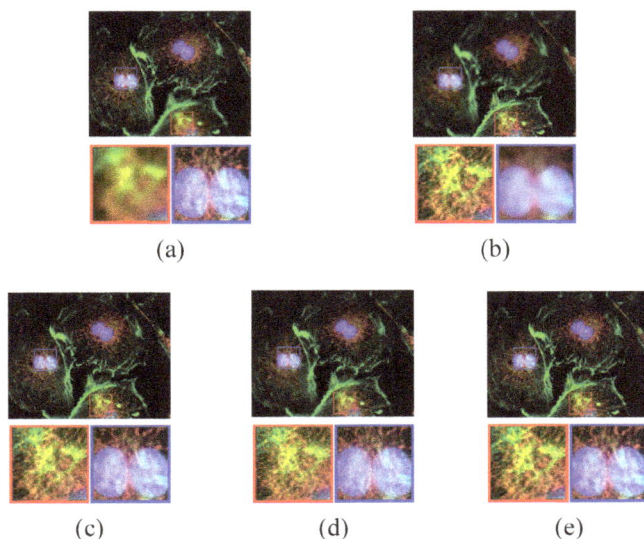

Figure 7. Fusion Results of Multi-focus Fluorescence Image - 2, (**a**,**b**) are source images, (**c**–**e**) are fused image of KSVD, JPDL, and proposed method respectively.

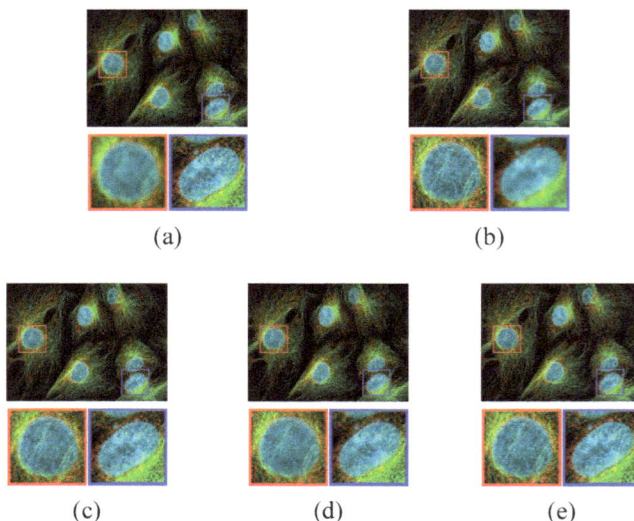

Figure 8. Fusion Results of Multi-focus Fluorescence Image - 3, (**a**,**b**) are source images, (**c–e**) are fused image of KSVD, JPDL, and proposed method respectively.

Objective metrics of multi-focus comparison experiment 2 and 3 are shown in Tables 2 and 3 respectively to evaluate the quality of fused images. According to the metric results, the proposed method has the best performance in all five objective evaluations in comparison experiment 2 and 3. So the proposed method has the best overall performance among all comparison methods.

Table 2. Fusion Performance Comparison - 2 of Multi-focus Fluorescence Image Pairs.

	$Q^{AB/F}$	MI	VIF	Q_Y	Q_{CB}
KSVD	0.5600	2.4747	0.6099	0.6161	0.5477
JPDL	0.6952	3.1357	0.7013	0.7355	0.7423
Proposed Solution	**0.7692***	**3.8982**	**0.7488**	**0.8058**	**0.8206**

* The highest result in each column is marked in bold-face.

Table 3. Fusion Performance Comparison - 3 of Multi-focus Fluorescence Image Pairs.

	$Q^{AB/F}$	MI	VIF	Q_Y	Q_{CB}
KSVD	0.6837	2.4624	0.8171	0.7855	0.6454
JPDL	0.8557	2.9358	0.8943	0.9325	0.7665
Proposed Solution	**0.8979***	**3.9154**	**0.9129**	**0.9790**	**0.8237**

* The highest result in each column is marked in bold-face.

3.3. Processing Time Comparison

Table 4 compares the processing times of three comparison experimentations. The proposed solution has lower computation costs than KSVD and JPDL in image fusion process. Compared with KSVD, the dictionary construction method of proposed solution does not use any iterative way to extract the underlying information of images, which is not efficient in dictionary construction. Although JPDL and the proposed method both cluster image pixels or patches based on geometric similarity, the proposed method does not use the iterative method of Steering Kernel Regression (SKR), which is time consuming.

Table 4. Processing Time Comparison.

	Experiment - 1	Experiment - 2	Experiment - 3
KSVD	164.02 s	215.09 s	108.65 s
JPDL	123.68 s	171.93 s	73.58 s
Proposed Solution	**76.23 s***	**103.96 s**	**25.47 s**

* The highest result in each column is marked in bold-face.

4. Conclusions

This paper proposes a novel sparse representation-based image fusion framework, which integrates geometric dictionary construction. A geometric image patch classification approach is presented to group image patches from different source images based on the similarity of image geometric structure. The proposed method extracts a few compact and informative sub-dictionaries from each image patch cluster by PCA and these sub-dictionaries are combined into a dictionary for sparse representation. Then image patches are sparsely coded into coefficients by the trained dictionary. For obtaining better edge and corner details of fusion results, the proposed solution also chooses image block size adaptively and selects optimal coefficients during the image process. The sparsely coded coefficients are fused by Max-L1 rule and inverted to the fused image. The proposed method is compared with existing mainstream sparse representation-based methods in various experiments. The experimentation results proves that the proposed method has good fusion performance in different image scenarios.

Acknowledgments: We would like to thank the supports by National Natural Science Foundation of China (61633005, 61403053).

Author Contributions: Zhiqin Zhu and Guanqiu Qi conceived and designed the experiments; Zhiqin Zhu and Guanqiu Qi performed the experiments; Zhiqin Zhu and Guanqiu Qi analyzed the data; Zhiqin Zhu contributed reagents/materials/analysis tools; Zhiqin Zhu and Guanqiu Qi wrote the paper; Yi Chai and Penghua Li provided technical support and revised the paper.

Conflicts of Interest: The authors declare no conflict of interest.

References

1. Tsai, W.; Qi, G. DICB: Dynamic Intelligent Customizable Benign Pricing Strategy for Cloud Computing. In Proceedings of the 5th IEEE International Conference on Cloud Computing, Honolulu, HI, USA, 24–29 June 2012; pp. 654–661.
2. Wu, W.; Tsai, W.; Jin, C.; Qi, G.; Luo, J. Test-Algebra Execution in a Cloud Environment. In Proceedings of the 8th IEEE International Symposium on Service Oriented System Engineering, SOSE 2014, Oxford, UK, 7–11 April 2014; pp. 59–69.
3. Tsai, W.; Qi, G.; Chen, Y. Choosing cost-effective configuration in cloud storage. In Proceedings of the 11th IEEE International Symposium on Autonomous Decentralized Systems, ISADS 2013, Mexico City, Mexico, 6–8 March 2013; pp. 1–8.
4. Li, X.; Li, H.; Yu, Z.; Kong, Y. Multifocus image fusion scheme based on the multiscale curvature in nonsubsampled contourlet transform domain. *Opt. Eng.* **2015**, *54*, 073115-1–073115-15.
5. Rodger, J.A. Toward reducing failure risk in an integrated vehicle health maintenance system: A fuzzy multi-sensor data fusion Kalman filter approach for {IVHMS}. *Expert Syst. Appl.* **2012**, *39*, 9821–9836.
6. Li, H.; Li, X.; Yu, Z.; Mao, C. Multifocus image fusion by combining with mixed-order structure tensors and multiscale neighborhood. *Inf. Sci.* **2016**, *349–350*, 25–49.
7. Sun, J.; Zheng, H.; Chai, Y.; Hu, Y.; Zhang, K.; Zhu, Z. A direct method for power system corrective control to relieve current violation in transient with UPFCs by barrier functions. *Int. J. Electr. Power & Energy Syst.* **2016**, *78*, 626–636.

8. Li, S.; Kang, X.; Hu, J.; Yang, B. Image matting for fusion of multi-focus images in dynamic scenes. *Inf. Fusion* **2013**, *14*, 147–162.
9. Li, H.; Liu, X.; Yu, Z.; Zhang, Y. Performance improvement scheme of multifocus image fusion derived by difference images. *Signal Process.* **2016**, *128*, 474–493.
10. Nejati, M.; Samavi, S.; Shirani, S. Multi-focus image fusion using dictionary-based sparse representation. *Inf. Fusion* **2015**, *25*, 72–84.
11. Li, S.; Yang, B. Multifocus image fusion using region segmentation and spatial frequency. *Image Vis. Comput.* **2008**, *26*, 971–979.
12. Li, H.; Yu, Z.; Mao, C. Fractional differential and variational method for image fusion and super-resolution. *Neurocomputing* **2016**, *171*, 138–148.
13. Li, H.; Qiu, H.; Yu, Z.; Zhang, Y. Infrared and visible image fusion scheme based on NSCT and low-level visual features. *Infrared Phys. Technol.* **2016**, *76*, 174–184.
14. Li, S.; Yang, B.; Hu, J. Performance comparison of different multi-resolution transforms for image fusion. *Inf. Fusion* **2011**, *12*, 74–84.
15. Vijayarajan, R.; Muttan, S. Discrete wavelet transform based principal component averaging fusion for medical images. *Int. J. Electron. Commun.* **2015**, *69*, 896–902.
16. Pajares, G.; de la Cruz, J.M. A wavelet-based image fusion tutorial. *Pattern Recognit.* **2004**, *37*, 1855–1872.
17. Makbol, N.M.; Khoo, B.E. Robust blind image watermarking scheme based on Redundant Discrete Wavelet Transform and Singular Value Decomposition. *Int. J. Electron. Commun.* **2013**, *67*, 102–112.
18. Luo, X.; Zhang, Z.; Wu, X. A novel algorithm of remote sensing image fusion based on shift-invariant Shearlet transform and regional selection. *Int. J. Electron. Commun.* **2016**, *70*, 186–197.
19. Liu, X.; Zhou, Y.; Wang, J. Image fusion based on shearlet transform and regional features. *Int. J. Electron. Commun.* **2014**, *68*, 471–477.
20. Sulochana, S.; Vidhya, R.; Manonmani, R. Optical image fusion using support value transform (SVT) and curvelets. *Optik-Int. J. Light Electron Opt.* **2015**, *126*, 1672–1675.
21. Zhu, Z.; Chai, Y.; Yin, H.; Li, Y.; Liu, Z. A novel dictionary learning approach for multi-modality medical image fusion. *Neurocomputing* **2016**, *214*, 471–482.
22. Seal, A.; Bhattacharjee, D.; Nasipuri, M. Human face recognition using random forest based fusion of à-trous wavelet transform coefficients from thermal and visible images. *Int. J. Electron. Commun.* **2016**, *70*, 1041–1049.
23. Qu, X.B.; Yan, J.W.; Xiao, H.Z.; Zhu, Z.Q. Image Fusion Algorithm Based on Spatial Frequency-Motivated Pulse Coupled Neural Networks in Nonsubsampled Contourlet Transform Domain. *Acta Autom. Sin.* **2008**, *34*, 1508–1514.
24. Tsai, W.; Qi, G. A Cloud-Based Platform for Crowdsourcing and Self-Organizing Learning. In Proceedings of the 8th IEEE International Symposium on Service Oriented System Engineering, SOSE 2014, Oxford, UK, 7–11 April 2014; pp.454–458.
25. Elad, M.; Aharon, M. Image Denoising Via Learned Dictionaries and Sparse representation. In Proceedings of the 2006 IEEE Computer Society Conference on Computer Vision and Pattern Recognition (CVPR 2006), New York, NY, USA, 17–22 June 2006; pp. 895–900.
26. Tsai, W.T.; Qi, G. Integrated fault detection and test algebra for combinatorial testing in TaaS (Testing-as-a-Service). *Simul. Model. Pract. Theory* **2016**, *68*, 108–124.
27. Zhu, Z.; Qi, G.; Chai, Y.; Yin, H.; Sun, J. A Novel Visible-infrared Image Fusion Framework for Smart City. *Int. J. Simul. Process Model.* **2016**, in press.
28. Han, J.; Yue, J.; Zhang, Y.; Bai, L. Local Sparse Structure Denoising for Low-Light-Level Image. *IEEE Trans. Image Process.* **2015**, *24*, 5177–5192.
29. Zhang, X.; Zhang, S. Diffusion scheme using mean filter and wavelet coefficient magnitude for image denoising. *Int. J. Electron. Commun.* **2016**, *70*, 944–952.
30. Sun, J.; Chai, Y.; Su, C.; Zhu, Z.; Luo, X. BLDC motor speed control system fault diagnosis based on LRGF neural network and adaptive lifting scheme. *Appl. Soft Comput.* **2014**, *14*, 609–622.
31. Xu, J.; Feng, A.; Hao, Y.; Zhang, X.; Han, Y. Image deblurring and denoising by an improved variational model. *Int. J. Electron. Commun.* **2016**, *70*, 1128–1133.
32. Shi, J.; Qi, C. Sparse modeling based image inpainting with local similarity constraint. In Proceedings of the IEEE International Conference on Image Processing, ICIP 2013, Melbourne, Australia, 15–18 September 2013; pp. 1371–1375.

33. Aslantas, V.; Bendes, E. A new image quality metric for image fusion: The sum of the correlations of differences. *Int. J. Electron. Commun.* **2015**, *69*, 1890–1896.
34. Yang, J.; Wright, J.; Huang, T.S.; Ma, Y. Image Super-Resolution Via Sparse Representation. *IEEE Trans. Image Process.* **2010**, *19*, 2861–2873.
35. Yang, B.; Li, S. Multifocus Image Fusion and Restoration With Sparse Representation. *IEEE Trans. Instrum. Meas.* **2010**, *59*, 884–892.
36. Li, H.; Li, L.; Zhang, J. Multi-focus image fusion based on sparse feature matrix decomposition and morphological filtering. *Opt. Commun.* **2015**, *342*, 1–11.
37. Wang, J.; Liu, H.; He, N. Exposure fusion based on sparse representation using approximate K-SVD. *Neurocomputing* **2014**, *135*, 145–154.
38. Kim, M.; Han, D.K.; Ko, H. Joint patch clustering-based dictionary learning for multimodal image fusion. *Inf. Fusion* **2016**, *27*, 198–214.
39. Yang, S.; Wang, M.; Chen, Y.; Sun, Y. Single-Image Super-Resolution Reconstruction via Learned Geometric Dictionaries and Clustered Sparse Coding. *IEEE Trans. Image Process.* **2012**, *21*, 4016–4028.
40. Sun, J.; Zheng, H.; Demarco, C.; Chai, Y. Energy Function Based Model Predictive Control with UPFCs for Relieving Power System Dynamic Current Violation *IEEE Trans. Smart Grid* **2016**, *7*, 2933–2942.
41. Keerqinhu; Qi, G.; Tsai, W.; Hong, Y.; Wang, W.; Hou, G.; Zhu, Z. Fault-Diagnosis for Reciprocating Compressors Using Big Data. In Proceedings of the Second IEEE International Conference on Big Data Computing Service and Applications, BigDataService 2016, Oxford, UK, 29 March–1 April 2016; pp. 72–81.
42. Bigün, J.; Granlund, G.H.; Wiklund, J. Multidimensional Orientation Estimation with Applications to Texture Analysis and Optical Flow. *IEEE Trans. Pattern Anal. Mach. Intell.* **1991**, *13*, 775–790.
43. Elad, M.; Yavneh, I. A Plurality of Sparse Representations is Better Than the Sparsest One Alone. *IEEE Trans. Inf. Theor.* **2009**, *55*, 4701–4714.
44. Dong, W.; Zhang, L.; Lukac, R.; Shi, G. Sparse Representation Based Image Interpolation With Nonlocal Autoregressive Modeling. *IEEE Trans. Image Process.* **2013**, *22*, 1382–1394.
45. Dong, W.; Zhang, L.; Shi, G.; Li, X. Nonlocally Centralized Sparse Representation for Image Restoration. *IEEE Trans. Image Process.* **2013**, *22*, 1620–1630.
46. Aharon, M.; Elad, M.; Bruckstein, A. *rmK*-SVD: An Algorithm for Designing Overcomplete Dictionaries for Sparse Representation. *IEEE Trans. Signal Process.* **2006**, *54*, 4311–4322.
47. Takeda, H.; Farsiu, S.; Milanfar, P. Kernel Regression for Image Processing and Reconstruction. *IEEE Trans. Image Process.* **2007**, *16*, 349–366.
48. Ratnarajah, T.; Vaillancourt, R.; Alvo, M. Eigenvalues and Condition Numbers of Complex Random Matrices. *SIAM J. Matrix Anal. Appl.* **2004**, *26*, 441–456.
49. Chatterjee, P.; Milanfar, P. Clustering-Based Denoising With Locally Learned Dictionaries. *IEEE Trans. Image Process.* **2009**, *18*, 1438–1451.
50. Yang, B.; Li, S. Pixel-level image fusion with simultaneous orthogonal matching pursuit. *Inf. Fusion* **2012**, *13*, 10–19.
51. Zhu, Z.; Qi, G.; Chai, Y.; Chen, Y. A Novel Multi-Focus Image Fusion Method Based on Stochastic Coordinate Coding and Local Density Peaks Clustering. *Future Internet* **2016**, *8*, 53.
52. Yin, H.; Li, S.; Fang, L. Simultaneous image fusion and super-resolution using sparse representation. *Inf. Fusion* **2013**, *14*, 229–240.
53. Petrovic, V.S. Subjective tests for image fusion evaluation and objective metric validation. *Inf. Fusion* **2007**, *8*, 208–216.
54. Tsai, W.; Colbourn, C.J.; Luo, J.; Qi, G.; Li, Q.; Bai, X. Test algebra for combinatorial testing. In Proceedings of the 8th IEEE International Workshop on Automation of Software Test, AST 2013, San Francisco, CA, USA, 18–19 May 2013; pp. 19–25.
55. Wang, Q.; Shen, Y.; Zhang, Y.; Zhang, J.Q. Fast quantitative correlation analysis and information deviation analysis for evaluating the performances of image fusion techniques. *IEEE Trans. Instrum. Meas.* **2004**, *53*, 1441–1447.
56. Sheikh, H.R.; Bovik, A.C. Image information and visual quality. *IEEE Trans. Image Process.* **2006**, *15*, 430–444.
57. Yang, C.; Zhang, J.Q.; Wang, X.R.; Liu, X. A novel similarity based quality metric for image fusion. *Inf. Fusion* **2008**, *9*, 156–160.

58. Liu, Z.; Blasch, E.; Xue, Z.; Zhao, J.; Laganiere, R.; Wu, W. Objective Assessment of Multiresolution Image Fusion Algorithms for Context Enhancement in Night Vision: A Comparative Study. *IEEE Trans. Pattern Anal. Mach. Intell.* **2011**, *34*, 94–109.

59. Chen, Y.; Blum, R.S. A new automated quality assessment algorithm for image fusion. *Image Vis. Comput.* **2009**, *27*, 1421–1432.

60. Tsai, W.T.; Qi, G.; Zhu, Z. Scalable SaaS Indexing Algorithms with Automated Redundancy and Recovery Management. *Int. J. Softw. Inform.* **2013**, *7*, 63–84.

61. Fluorescence Image Example - 1. Available online: https://www.thermofisher.com/cn/zh/home/references /newsletters- and-journals/probesonline/probesonline-issues-2014/probesonline-jan-2014.html/ (accessed on 21 December 2016).

62. Fluorescence Image Example - 2. Available online: http://www2.warwick.ac.uk/services/ris/ impactinnovation/impact/analyticalguide/fluorescence/ (accessed on 21 December 2016).

63. Fluorescence Image Example - 3. Available online: http://www.lichtstadt-jena.de/erfolgsgeschichten/ themen-detailseite/ (accessed on 21 December 2016).

applied
sciences

MDPI

Article

Infrared Spectroscopy as Molecular Probe of the Macroscopic Metal-Liquid Interface

Johannes Kiefer [1,2,3,4,*], Johan Zetterberg [5], Andreas Ehn [5], Jonas Evertsson [6], Gary Harlow [6] and Edvin Lundgren [6]

1 Technische Thermodynamik, University of Bremen, Badgasteiner Str. 1, 28359 Bremen, Germany
2 School of Engineering, University of Aberdeen, Aberdeen AB24 3UE, UK
3 Erlangen Graduate School in Advanced Optical Technologies (SAOT), Universität Erlangen-Nürnberg, 91052 Erlangen, Germany
4 MAPEX Center for Materials and Processes, University of Bremen, Bibliothekstr. 1, 28359 Bremen, Germany
5 Combustion Physics, Lund University, P.O. Box 118, 221 00 Lund, Sweden; johan.zetterberg@forbrf.lth.se (J.Z.); andreas.ehn@forbrf.lth.se (A.E.)
6 Division of Synchrotron Radiation Research, Lund University, 221 00 Lund, Sweden; jonas.evertsson@sljus.lu.se (J.E.); gary.harlow@sljus.lu.se (G.H.); edvin.lundgren@sljus.lu.se (E.L.)
* Correspondence: jkiefer@uni-bremen.de; Tel.: +49-421-218-64-777

Received: 7 October 2017; Accepted: 23 November 2017; Published: 28 November 2017

Abstract: Metal-liquid interfaces are of the utmost importance in a number of scientific areas, including electrochemistry and catalysis. However, complicated analytical methods and sample preparation are usually required to study the interfacial phenomena. We propose an infrared spectroscopic approach that enables investigating the molecular interactions at the interface, but needing only minimal or no sample preparation. For this purpose, the internal reflection element (IRE) is wetted with a solution as first step. Second, a small plate of the metal of interest is put on top and pressed onto the IRE. The tiny amount of liquid that is remaining between the IRE and the metal is sufficient to produce an IR spectrum with good signal to noise ratio, from which information about molecular interactions, such as hydrogen bonding, can be deduced. Proof-of-concept experiments were carried out with aqueous salt and acid solutions and an aluminum plate.

Keywords: hydrogen bonding; ATR-FTIR; adsorption

1. Introduction

Interfaces between metals and fluids are omnipresent. Electrochemistry is the most obvious area, when an electrolyte is in contact with an electrode. This is not only the case in electrocatalysis [1,2] and energy storage devices like in a battery [3,4], but also when metals are undergoing corrosion [5,6]. Despite the importance of these systems, the analysis of the underlying physical and chemical phenomena is still a challenge. This is particularly true for those processes that are happening directly at the interface.

In a reacting system, e.g., when an electrochemical cell is under operation, an integral piece of information can be obtained by monitoring the electrical current and voltage, and/or by analyzing the bulk fluid at some distance to the surface. This allows for deducing information about the overall chemical oxidation and reduction reactions [7]. However, details about the molecular interactions at the surface are more difficult to investigate. Moreover, when the fluid (or some of its constituents) and the metal are not reacting, no information can be provided by measurements of current and bulk diagnostics.

Non-reacting metal-fluid interfaces call for surface sensitive methods. Unfortunately, the number of possible analytical techniques is rather limited. X-ray and ultraviolet radiation based methods, such as photoelectron spectroscopy, usually require vacuum conditions and hence their application to samples containing volatile liquids is difficult [8]. Scattering and diffraction methods are usually limited to crystalline materials, and often even to atomically flat single crystal surfaces. Suitable approaches, on the other hand, can be found in the vibrational spectroscopy toolbox. Reflection-absorption infrared spectroscopy can be used to study heterogeneous catalytic processes, e.g., to identify intermediate species at the gas-solid interface [9,10]. However, its application to aqueous systems, e.g., a salt solution on a metal surface, is difficult due to the strong absorption of water in the mid-infrared.

Second-order (or even-order, in general) nonlinear effects are advantageous for studying surfaces and interfaces [11]. The even-order susceptibility is zero in the bulk of a fluid, and hence only the molecules in the optically anisotropic interfacial layer contribute to the signal, in particular when the molecules at the interface are oriented in a certain manner. In other words, the signal of even-order methods is highly surface-specific, while methods utilizing odd-order effects like infrared absorption and Raman scattering may be biased by signals from the bulk. Second-order vibrational spectroscopy can, for instance, be performed in terms of sum-frequency generation (SFG) [11,12]. However, when the IR radiation must travel through a highly absorbing medium like water, the resulting signal levels may be low. The strong absorption of water can be avoided in Raman spectroscopy because the excitation wavelength may basically be chosen arbitrarily [13], and thus it can be spectrally separated from any absorption bands. Consequently, surface-enhanced Raman scattering (SERS) can be a solution if the metal is capable of coupling with the electric field of an incident laser beam to result in the plasmonic enhancement of the inherently weak Raman signal [14]. However, the application of SERS and the data evaluation is not straightforward due to a limited reproducibility.

As a possible approach that can overcome all the above mentioned difficulties, we propose the use of attenuated total reflection infrared (ATR-IR) spectroscopy. ATR techniques are generally advantageous for studying highly absorbing media, as the radiation interacts with the sample only in an evanescent field, and hence the attenuation of the intensity is moderate [15,16]. ATR-IR has proven its potential for studying molecular phenomena at liquid-liquid and liquid-solid interfaces, including aqueous systems in the past [17–19]. However, all of the methods that are proposed to date share the disadvantage of requiring specialized equipment and/or sample preparation. For example, building an electrochemical cell on top of the spectrometer or depositing particles or a thin solid film on the ATR crystal are frequently employed approaches [20–22]. Recently, Koichumanova et al. [23] studied metal-liquid interfaces by an ATR-IR approach. They immobilized a catalyst material directly on the surface of the internal reflection element (IRE). A similar method was applied by Mundunkatowa et al. [24]. Kraack et al. [25,26] demonstrated a surface-enhanced ATR technique, utilizing gold or platinum nanoparticles at the IRE. Aguirre and co-workers [27] presented spectra from the solid-liquid interface in a specially designed microfluidic reactor. In summary, in the past, all of the work on solid-liquid interfaces using ATR-IR spectroscopy focused on adsorption directly at the IRE, adsorption at coated IREs, or adsorption to particulate matter in contact with the IRE [28].

2. Materials and Methods

The present study proposes a more straightforward technique for analyzing the liquid-solid interface. For the purpose of analyzing the molecular interactions at the interface of a metal and an aqueous solution, a modified version of the recently proposed solvent infrared spectroscopy (SIRS) method seems most suitable [18]. SIRS was developed to study the surface chemistry of nanopowders utilizing the influence of the functional groups at the particle surface on the vibrational structure of a solvent. In a typical SIRS experiment, the nanopowder was first pressed onto the internal reflection element of an ATR-IR instrument. In the second step, a solvent, such as water or an alcohol, was added to fill the void space in the fixed bed of particles. The comparison of the solvent spectrum with and without the particles yields information about the molecular interactions at and the chemistry of the surface [18].

This approach can be modified to study the molecular phenomena at a macroscopic metal-liquid interface. Figure 1 illustrates the proposed modified SIRS technique. Here, the first step is to add a droplet of the fluid onto the internal reflection element. Thereafter, the metal plate is placed on top of the fluid and pressed down by the stamp of the instrument. When the stamp is applying a force on the metal plate, most of the fluid will flow away from the IRE. Eventually, only a very thin film will remain due to the surface roughness of the metal plate. This is beneficial as there will be virtually no bulk fluid left, but the IRE will still be wetted. Consequently, the ATR-IR spectrum will carry information about the molecular interactions at the surface. As a significant advantage, the described technique requires virtually no sample preparation. This is particularly true in comparison to the approaches that are reported in the literature, which were all based on the deposition of a thin film on the IRE.

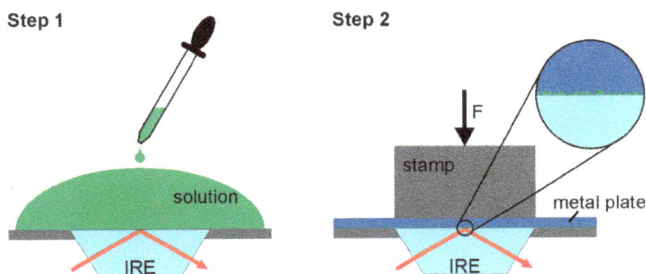

Figure 1. Schematic of the experimental procedure. IRE = internal reflection element; F = force.

As proof of the concept, experiments of three aqueous solutions (pure water, sodium sulfate solution, citric acid solution) at an aluminum surface were carried out on an Agilent Cary 630 instrument equipped with a diamond ATR unit (1 reflection, 2 cm^{-1} resolution). Deionized water (18.2 MΩ cm) was provided from a Millipore device. Aqueous solutions of 2.1 mol % Na_2SO_4 and 2.8 mol % citric acid were prepared gravimetrically.

3. Results

The pairs of spectra recorded with and without an aluminum plate are displayed in Figure 2. Note that the spectral window between 1800 and 2800 cm^{-1} does not contain any appreciable signals, and was therefore omitted. The broad and strong hydroxyl (OH) stretching band of water dominates the high wavenumber region, i.e., 2800 to 3800 cm^{-1}. This band is commonly deconvolved into several sub-bands, indicating the different hydrogen-bonding states of water molecules [29,30]: The lower the wavenumber, the stronger the hydrogen bonding [31]. However, even with the naked eye, systematic differences between the three cases can be observed. Pure water (without Al plate) has the strongest hydrogen bonding (HB) network, as indicated by a maximum at the low wavenumber side of the band. The citric acid solution has the weakest HB network, and the sodium sulfate solution is in between the two. The fingerprint region of the spectra shows a common peak between 1630 and 1640 cm^{-1}, owing to the OH bending vibration of water. The spectrum of the sulfate solution additionally exhibits a strong S = O stretching band at 1091 cm^{-1}. The citric acid solution shows a multitude of peaks in the fingerprint region. The most dominant ones appear at 1713 and 1222 cm^{-1}, and can be assigned to the carboxylic acid groups [32].

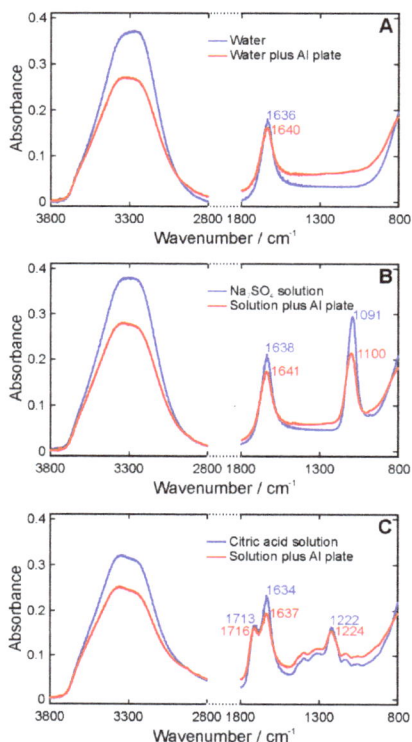

Figure 2. Fingerprint and hydroxyl (OH) stretching region of the IR spectra recorded with and without aluminum plate. (**A**): water; (**B**): aqueous sodium sulfate solution; and, (**C**): aqueous citric acid solution.

When the Al plate is added, all three spectra change significantly. The intensity of the OH stretching band is reduced by about 20–30%. At first glance, this reduction appears small when we argue that we move from studying the bulk fluid to the interface. However, the working principle of ATR needs to be kept in mind here. The penetration depth d_p of the evanescent field is commonly described as

$$d_p = \frac{\lambda}{2\pi \cdot n_{IRE} \cdot \left(\sin^2 \alpha - \left(\frac{n_{sample}}{n_{IRE}} \right)^2 \right)^{1/2}} \tag{1}$$

with the wavelength λ, the reflection angle α, and the refractive indices of the IRE and the sample, n_{IRE} and n_{sample}, respectively. When we assume that the liquid wets the diamond surface even in the presence of the Al plate, this penetration depth will remain the same. So, the key is to get the metal so close to the IRE surface that the probed molecules represent the thin interfacial layer. This is why a force needs to be applied to the metal plate in order to press it onto the IRE. The effective path length, on the other hand, determines the absolute absorbance, but it is an auxiliary parameter that cannot be measured. It represents the absorption path that would be necessary in a transmission experiment to observe the same attenuation of the radiation [17,33].

The shape of the OH stretching band alters such that the high-wavenumber wing becomes dominant in all three cases. Figure 3 illustrates the absorbance normalized bands and their differences in order to emphasize this effect. This change suggests a weakening of the HB network in the presence of the Al surface. The Al plate exhibits a thin oxide layer and the O^{2-} ions can act as

HB acceptors. Moreover, the Al^{3+} ions can interact with water oxygen atoms via Coulomb forces. However, the interactions at the surface are weaker than the hydrogen bonds in fully tetrahedrally coordinated bulk water. The change in the molecular interactions is further supported by the peaks in the fingerprint region. The OH bending mode slightly shifts towards higher wavenumber. The same behavior can be noted for the sulfate and the carboxylic acid peaks.

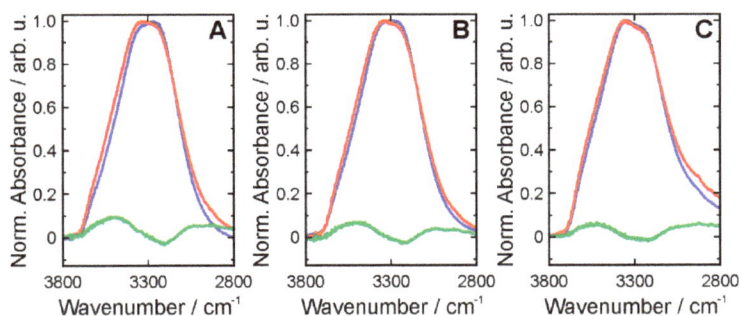

Figure 3. Normalized spectra (color code as in Figure 2) and difference spectra (green) in the OH stretching region. (**A**): water; (**B**): aqueous sodium sulfate solution; and, (**C**): aqueous citric acid solution.

4. Conclusions

In conclusion, we have shown that ATR-IR spectroscopy is capable of analyzing the molecular interactions between solvent molecules and a metal surface. For this purpose, the solvent or a solution was initially placed on top of the ATR crystal, and, in the second step, a metal plate was added and pressed onto the crystal. The fluid remaining in a very thin layer between the ATR crystal and the metal plate allows for recording the IR spectrum of the interfacial molecules. The IR spectrum yields information about the intermolecular interactions at the surface via monitoring of their effects on the vibrational structure of the solvent and solute. However, it needs to be kept in mind that the signals do not exclusively originate from the interfacial mono-molecular layer, but from a thin layer that is determined by the penetration depth of the ATR approach. Applying a force to the metal plate ensures that the fluid layer is thin and that the interfacial molecular interactions are sufficiently prominent in the spectra.

Using the example of water and aqueous solutions of sodium sulfate and citric acid in contact with an aluminum plate, it was shown that the presence of an Al surface results in a weakening of the hydrogen bonding network between the water molecules. A key advantage of the proposed method is its experimental simplicity and that no specialized or custom-made equipment is needed. Measurements can be performed on any ATR-IR instrument in a straightforward manner.

Author Contributions: J.Z., A.E. and E.L. initiated the project; J.K., J.E., J.Z., A.E. and E.L. conceived and designed the experiments; J.K. performed the experiments; J.K. analyzed the data; J.E., G.H. and E.L. contributed reagents/materials/analysis tools; all authors contributed to the interpretation of the data; all authors contributed to writing the paper with J.K. preparing the first draft.

Conflicts of Interest: The authors declare no conflict of interest.

References

1. Seh, Z.W.; Kibsgaard, J.; Dickens, C.F.; Chorkendorff, I.B.; Norskov, J.K.; Jaramillo, T.F. Combining theory and experiment in electrocatalysis: Insights into materials design. *Science* **2017**, *335*, 146. [CrossRef] [PubMed]
2. Stamenkovic, V.R.; Strmcnik, D.; Lopes, P.P.; Markovic, N.M. Energy and fuels from electrochemical interfaces. *Nat. Mater.* **2017**, *16*, 57–69. [CrossRef] [PubMed]

3. Wang, H.; Yang, Y.; Guo, L. Nature-Inspired Electrochemical Energy-Storage Materials and Devices. *Adv. Energy Mater.* **2017**, *7*, 1601709. [CrossRef]
4. Cho, J.; Jeong, S.; Kim, Y. Commercial and research battery technologies for electrical energy storage applications. *Prog. Energy Combust. Sci.* **2015**, *48*, 84–101. [CrossRef]
5. Manam, N.S.; Harun, W.S.W.; Shri, D.N.A.; Ghani, S.A.C.; Kurniawan, T.; Ismail, M.H.; Ibrahim, M.H.I. Study of corrosion in biocompatible metals for implants: A review. *J. Alloys Compd.* **2017**, *701*, 698–715. [CrossRef]
6. Dwivedi, D.; Lepkova, K.; Becker, T. Carbon steel corrosion: A review of key surface properties and characterization methods. *RSC Adv.* **2017**, *7*, 4580–4610. [CrossRef]
7. Zehentbauer, F.M.; Bain, E.J.; Kiefer, J. Multiple parameter monitoring in a direct methanol fuel cell. *Meas. Sci. Technol.* **2012**, *23*, 045602. [CrossRef]
8. Salmeron, M.; Schlögl, R. Ambient pressure photoelectron spectroscopy: A new tool for surface science and nanotechnology. *Surf. Sci. Rep.* **2008**, *63*, 169–199. [CrossRef]
9. Rasko, J.; Domok, A.; Baan, K.; Erdohelyi, A. FTIR and mass spectrometric study of the interaction of ethanol and ethanol-water with oxide-supported platinum catalysts. *Appl. Catal. A* **2006**, *299*, 202–211. [CrossRef]
10. McCue, A.J.; Mutch, G.A.; McNab, A.I.; Campbell, S.; Anderson, J.A. Quantitative determination of surface species and adsorption sites using Infrared spectroscopy. *Catal. Today* **2016**, *259*, 19–26. [CrossRef]
11. Kraack, J.P.; Hamm, P. Surface-sensitive and surface-specific ultrafast two-dimensional vibrational spectroscopy. *Chem. Rev.* **2017**, *117*, 10623–10664. [CrossRef] [PubMed]
12. Vidal, F.; Tadjeddine, A. Sum-frequency generation spectroscopy of interfaces. *Rep. Prog. Phys.* **2005**, *68*, 1095–1127. [CrossRef]
13. Kiefer, J. Recent advances in the characterization of gaseous and liquid fuels by vibrational spectroscopy. *Energies* **2015**, *8*, 3165–3197. [CrossRef]
14. McNay, G.; Eustace, D.; Smith, W.E.; Faulds, K.; Graham, D. Surface-Enhanced Raman Scattering (SERS) and Surface-Enhanced Resonance Raman Scattering (SERRS): A Review of Applications. *Appl. Spectrosc.* **2011**, *65*, 825–837. [CrossRef] [PubMed]
15. Chalmers, J.M.; Griffiths, P.R. *Handbook of Vibrational Spectroscopy*; John Wiley & Sons: Hoboken, NJ, USA, 2001.
16. Kiefer, J. Simultaneous acquisition of absorption and fluorescence spectra of strong absorbers utilizing an evanescent supercontinuum. *Opt. Lett.* **2016**, *41*, 5684–5687. [CrossRef] [PubMed]
17. Kiefer, J.; Frank, K.; Schuchmann, H.P. Attenuated total reflection infrared (ATR-IR) spectroscopy of a water-in-oil emulsion. *Appl. Spectrosc.* **2011**, *65*, 1024–1028. [CrossRef] [PubMed]
18. Kiefer, J.; Grabow, J.; Kurland, H.-D.; Müller, F.A. Characterization of Nanoparticles by Solvent Infrared Spectroscopy. *Anal. Chem.* **2015**, *87*, 12313–12317. [CrossRef] [PubMed]
19. Torregrosa-Coque, R.; Alvarez-Garcia, S.; Martin-Martinez, J.M. Migration of low molecular weight moiety at rubber-polyurethane interface: An ATR-IR spectroscopy study. *Int. J. Adhes. Adhes.* **2011**, *31*, 389–397. [CrossRef]
20. Wang, H.; Zhou, Y.W.; Cai, W.B. Recent applications of in situ ATR-IR spectroscopy in interfacial electrochemistry. *Curr. Opin. Electrochem.* **2017**, *1*, 73–79. [CrossRef]
21. Zandi, O.; Hamann, T.W. Determination of photoelectrochemical water oxidation intermediates on haematite electrode surfaces using operando infrared spectroscopy. *Nat. Chem.* **2016**, *8*, 778–783. [CrossRef] [PubMed]
22. Osawa, M. Dynamic process in electrochemical reactions studied by surface-enhanced infrared absorption spectroscopy (SEIRAS). *Bull. Chem. Soc. Jpn.* **1997**, *70*, 2861–2880. [CrossRef]
23. Koichumanova, K.; Visan, A.; Geerdink, B.; Lammertink, R.G.H.; Mojet, B.L.; Seshan, K.; Lefferts, L. ATR-IR spectroscopic cell for in situ studies at solid-liquid interface at elevated temperatures and pressures. *Catal. Today* **2017**, *283*, 185–194. [CrossRef]
24. Mudunkotuwa, I.A.; Al Minshid, A.; Grassian, V.H. ATR-FTIR spectroscopy as a tool to probe surface adsorption on nanoparticles at the liquid-solid interface in environmentally and biologically relevant media. *Analyst* **2014**, *139*, 870–881. [CrossRef] [PubMed]
25. Kraack, J.P.; Kaech, A.; Hamm, P. Surface Enhancement in Ultrafast 2D ATR IR Spectroscopy at the Metal-Liquid Interface. *J. Phys. Chem. C* **2016**, *120*, 3350–3359. [CrossRef]
26. Kraack, J.P.; Kaech, A.; Hamm, P. Molecula-specific interactions of diatomic adsorbates at metal-liquid interfaces. *Struct. Dyn.* **2017**, *4*, 044009. [CrossRef] [PubMed]

27.	Aguirre, A.; Kler, P.A.; Berli, C.L.A.; Collins, S.E. Design and operational limits of an ATR-FTIR spectroscopic microreactor for investigating reactions at liquid-solid interface. *Chem. Eng. J.* **2014**, *243*, 197–206. [CrossRef]

28.	Hind, A.R.; Bhargava, S.K.; McKinnon, A. At the solid/liquid interface: FTIR/ATR—The tool of choice. *Adv. Colloid Interface Sci.* **2001**, *93*, 91–114. [CrossRef]

29.	Schmidt, D.A.; Miki, K. Structural correlations in liquid water: A new interpretation of IR spectroscopy. *J. Phys. Chem. A* **2007**, *111*, 10119–10122. [CrossRef] [PubMed]

30.	Wallace, V.M.; Dhumal, N.R.; Zehentbauer, F.M.; Kim, H.J.; Kiefer, J. Revisiting the Aqueous Solutions of Dimethyl Sulfoxide by Spectroscopy in the Mid- and Near-Infrared: Experiments and Car-Parrinello Simulations. *J. Phys. Chem. B* **2015**, *119*, 14780–14789. [CrossRef] [PubMed]

31.	Joseph, J.; Jemmis, E.D. Red-, blue-, or no-shift in hydrogen bonds: A unified explanation. *J. Am. Chem. Soc.* **2007**, *129*, 4620–4632. [CrossRef] [PubMed]

32.	Dhumal, N.R.; Singh, M.P.; Anderson, J.A.; Kiefer, J.; Kim, H.J. Molecular interactions of a Cu-based metal-organic framework with a confined imidazolium-based ionic liquid: A combined density functional theory and experimental vibrational spectroscopy study. *J. Phys. Chem. C* **2016**, *120*, 3295–3304. [CrossRef]

33.	Averett, L.A.; Griffiths, P.R. Effective path length in attenuated total reflection spectroscopy. *Anal. Chem.* **2008**, *80*, 3045–3049. [CrossRef] [PubMed]

![applied sciences logo] *applied sciences*

MDPI

Article

Probing Structures of Interfacial 1-Butyl-3-Methylimidazolium Trifluoromethanesulfonate Ionic Liquid on Nano-Aluminum Oxide Surfaces Using High-Pressure Infrared Spectroscopy

Hai-Chou Chang [1,*], Teng-Hui Wang [1] and Christopher M. Burba [2]

[1] Department of Chemistry, National Dong Hwa University, Shoufeng, Hualien 974, Taiwan; 410112029@ems.ndhu.edu.tw

[2] Department of Natural Science, Northeastern State University, Tahlequah, OK 74464, USA; burba@nsuok.edu

* Correspondence: hcchang@gms.ndhu.edu.tw; Tel.: +886-3-863-3585

Received: 20 July 2017; Accepted: 15 August 2017; Published: 18 August 2017

Abstract: The interactions between 1-butyl-3-methylimidazolium trifluoromethanesulfonate ([BMIM][TFS]) and nano-Al_2O_3 are studied using high-pressure infrared spectroscopy. The thickness of the [BMIM][TFS] interfacial layer on the aluminum oxide are adjusted by controlling the number of washes with ethanol. In contrast to the results obtained under ambient pressure, local structures of both the cations and anions of [BMIM][TFS] are disturbed under high pressures. For example, bands due to C-H stretching motions display remarkable blue-shifts in frequency as the pressure of the [BMIM][TFS]/Al_2O_3 composites is increased to 0.4 GPa. The bands then undergo mild shifts in frequency upon further compression. The discontinuous jump occurring around 0.4 GPa becomes less obvious when the amount of ionic liquid on the Al_2O_3 is reduced by washing with ethanol. The nano-Al_2O_3 with surfaces may weaken the cation/anion interactions in the interfacial area as a result of the formation of pressure-enhanced Al_2O_3/ionic liquid interactions under high pressures.

Keywords: imidazolium-based ionic liquids; infrared (IR); spectroscopy; high pressures

1. Introduction

Metal oxide nanoparticles possess different physicochemical properties in comparison to bulk materials, and the size of nanoparticles is responsible for the changes in their characteristics. Aluminum oxide, commonly called alumina, has tremendous applications in ceramics, medical products, and catalysis, to name a few [1–6]. Despite the expanding number of applications for aluminum oxide, the physical arrangement and interactions of liquids at Al_2O_3 surfaces remains poorly understood. Many experimental and theoretical studies have indicated the inherent complexity of interfacial interactions among certain liquid/Al_2O_3 mixtures, and the immobilization of liquids on Al_2O_3 surfaces may improve the applicability of these materials in industry [4–6].

Ionic liquids (ILs) are a subset of low-melting-point salts that are known for high ionic conductivities, low vapor pressures, and low flammability. This combination of properties has opened opportunities for ionic liquids to serve as alternatives for conventional volatile organic solvents [7–10]. Although ILs offer advantages over conventional solvents, there is still relatively little information available on the structures of ILs in the bulk and along various interfaces [7,9]. The most extensively studied ILs are 1-alkyl-3-methylimidazolium salts, which feature asymmetric cations [7–10]. The imidazolium ions are usually paired with anions that have highly delocalized charges or are intrinsically bulky. The liquid structures of 1-alkyl-3-methylimidazolium salts are strongly affected by the long-range charge organization of the constituent ions and may be described in terms of an extended quasi-lattice, in which ions are viewed as occupying sites in a disorganized crystalline lattice [11,12]. In this model, any particular ion may be viewed as being trapped within a "cage" that is defined by its nearest neighbors. For ILs, the immediate solvation shell of an ion is most likely to be populated with counterions. Thus, an ion will have frequent opportunities to form short-lived associated species (e.g., ionic pairs, clusters, or aggregates) with the ions composing its cage [7,13]. With three hydrogen atoms bound to the imidazolium ring, the most acidic proton $(C^2\text{-H})$ may form cation/anion hydrogen bonds. The nonpolar alkyl side chain of the cation provides another source of mesoscopic self-assembly of imidazolium ILs into nanostructured polar/nonpolar domains. Such nano-segregation is frequently observed in small angle X-ray scattering experiments for alkyl side chains containing four or more carbon atoms [14]. Various studies demonstrate significant heterogeneities in certain ionic liquid dynamics, and much of the heterogeneity may arise from nano-structured associations with polar and nonpolar regions [7–10,13,14].

Microscopic structures of certain ILs at solid surfaces have received considerable attention in recent years [7,9,13]. Nevertheless, the interfacial interactions caused by ionic liquid adsorption on the surface of Al_2O_3 are complex and depend on the molecular structures of ILs. It has been suggested that Coulombic forces, hydrogen bonding, and π interactions may each play significant roles in defining the molecular ordering of ILs along the surface [7,9]. In this study, 1-butyl-3-methylimidazolium trifluoromethanesulfonate [BMIM][TFS] and nano-Al_2O_3 were chosen as a model IL and solid support, respectively. As pointed out by Andanson et al. [4], the TFS anion has many strong bands in the infrared spectrum. Unlike the bis(trifluoromethylsulfonyl)amide $(NTf_2{}^-)$ anion, the TFS anion has only one conformer, which simplifies the spectral analysis. Nano-Al_2O_3 was chosen in this study mainly due to the wide application of nano-Al_2O_3, particularly in the field of catalysis.

Vibrational spectroscopy is particularly useful in studying ionic liquid/Al_2O_3 interactions because the frequencies and intensities of cation and anion vibrational modes are sensitive to their immediate potential energy environments. In this way, changes in band frequencies and intensities may be used to infer changes in the local structures of the ions. Many spectroscopic studies on ionic liquid interactions with nanostructured metal oxides focus on changing the temperature of the composite or varying the molecular structure of the ionic liquid in contact with the solid (e.g., lengthening the alkyl side chain) to probe the interfacial interactions [1–9]. Hydrostatic pressure is another degree of freedom that can continuously tune the structures, interactions, and solvation of chemical systems. In the case of temperature changes, the sample's properties can change significantly due to a simultaneous change in thermal energy and volume. However, the use of pressure as a variable allows one to separate the thermal and volume effects. Studies have shown the potential significance that pressure has on controlling the strength of intermolecular hydrogen bonding, especially the relatively weak C-H—X hydrogen bonds, where X is an anion atom [15–18]. Pressure-dependent changes in spectral features indicate that nanoparticles tend to influence local structures of certain ILs under high pressures. As certain ILs have been employed as graphene stabilizers, previous studies suggest that imidazolium ILs with short alkyl chain lengths ($n < 4$) may be suitable choices to modulate the performance of energy storage devices via pressure-enhanced interfacial interactions [15]. In order to obtain detailed insights into the ionic liquid/Al_2O_3 interactions, we use variable pressure as a window into the nature of hydrogen bonding structures of imidazolium ionic liquid/Al_2O_3 mixtures in this article.

2. Materials and Methods

Samples were prepared using 1-butyl-3-methylimidazolium trifluoromethanesulfonate ([BMIM][TFS], >95%, Fluka, Morris Plains, NJ, USA), and nanosized-aluminum oxide (nano-Al_2O_3, <50 nm, Aldrich, St. Louis, MO, USA). The nano-Al_2O_3 powders (ca. 0.006 g) were mixed with [BMIM][TFS] (ca. 0.06 g), followed by sonication for 30 min, centrifugation, several washes of the precipitate with ethanol (0.1 mL for a single wash), centrifugation, and drying under vacuum for

3 h. The infrared spectra of samples measured at ambient pressure were taken by filling samples in a cell characterized by two CaF$_2$ windows without the spacers. A diamond anvil cell (DAC) of Merrill-Bassett design with a diamond culet size of 0.6 mm was used to generate pressures of up to ca. 2 GPa. Two type-IIa diamonds were used for mid-infrared measurements. Infrared (IR) spectra of the samples were measured on a Fourier transform spectrophotometer (Spectrum RXI, Perkin-Elmer, Naperville, IL, USA) equipped with an LITA (lithium tantalite) mid-infrared detector. The infrared beam was condensed through a 5× beam condenser onto the sample in the diamond anvil cell. To remove the absorption of the diamond anvils, the absorption spectra of the DAC were measured first and subtracted from those of the samples. Samples were contained in a 0.3-mm-diameter hole in a 0.25-mm-thick Inconel gasket mounted on the DAC. To reduce the absorbance of the samples, CaF$_2$ crystals (prepared from a CaF$_2$ optical window) were placed into the holes and compressed to be transparent prior to inserting the samples. A sample droplet filled the empty space of the entire hole of the gasket in the DAC, which was subsequently sealed when the opposing anvils were pushed toward one another. Typically, we chose a resolution of 4 cm^{-1} (data point resolution of 2 cm^{-1}). For each spectrum, 1000 scans were compiled. Pressure calibration was performed following Wong's method [19,20].

3. Results and Discussion

Figure 1 shows the infrared spectra of pure [BMIM][TFS], the [BMIM][TFS]/Al$_2$O$_3$ mixture, the mixture after two washes with ethanol, and the mixture after five washes obtained under ambient pressure. The amount of [BMIM][TFS] are adjusted by controlling the numbers of washes in this study; the amount of [BMIM][TFS] becomes almost depleted after five washes, as shown in Figure 1d. The IR spectrum of pure [BMIM][TFS] in Figure 1a shows two imidazolium C-H bands at 3116 and 3155 cm^{-1}, a shoulder to the 3155 cm^{-1} band at 3098 cm^{-1}, and three alkyl C-H bands at 2877, 2940, and 2967 cm^{-1}, respectively [21]. As shown in Figure 1, no appreciable changes in spectral features and band frequencies of C-H vibrations occurred as [BMIM][TFS] was mixed with nano-Al$_2$O$_3$, except for a noticeable absorption caused by O-H groups from the surfaces of nano-Al$_2$O$_3$ that appears in Figure 1d. These results indicate that IR measurements recorded under ambient pressure may not be sensitive enough to monitor the interfacial interactions between the BMIM$^+$ cation and nano-Al$_2$O$_3$.

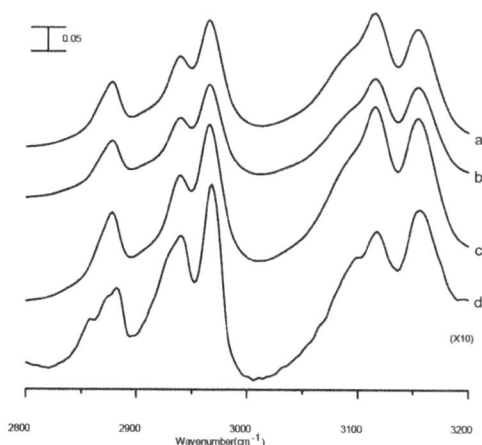

Figure 1. Infrared (IR) spectra of pure 1-butyl-3-methylimidazolium trifluoromethanesulfonate ([BMIM][TFS]) (curve a), the [BMIM][TFS]/Al$_2$O$_3$ mixture (curve b), the mixture after two washes with ethanol (curve c), and the mixture after five washes (curve d), recorded under ambient pressure. Curve d was baseline-corrected by subtracting a straight line. Relative intensities of the IR bands are given in increments of 0.05; the scale bar is displayed in the upper left corner of the figure.

Figure 2 displays the IR spectra of samples in the spectral range between 1000 and 1350 cm^{-1}. This region of the spectrum is predominately dominated by absorption bands due to vibrations of the TFS$^-$ anions. The bands at ca. 1165 and 1226 cm^{-1} are assigned to CF$_3$ asymmetric and symmetric stretching vibrations, respectively. The absorption bands located near 1032 and 1270 cm^{-1} (doublet) are assigned to the symmetric and asymmetric stretching vibration of the SO$_3$ group of the TFS$^-$ anion, respectively. Comparing the spectral features of Figure 2d with those of Figure 2a reveal mild frequency shifts and narrowing for the bands at ca. 1165 and 1270 cm^{-1}, caused by the presence of nano-Al$_2$O$_3$. The absorption of CaF$_2$ windows (below 1050 cm^{-1}) also becomes obvious in Figure 2d.

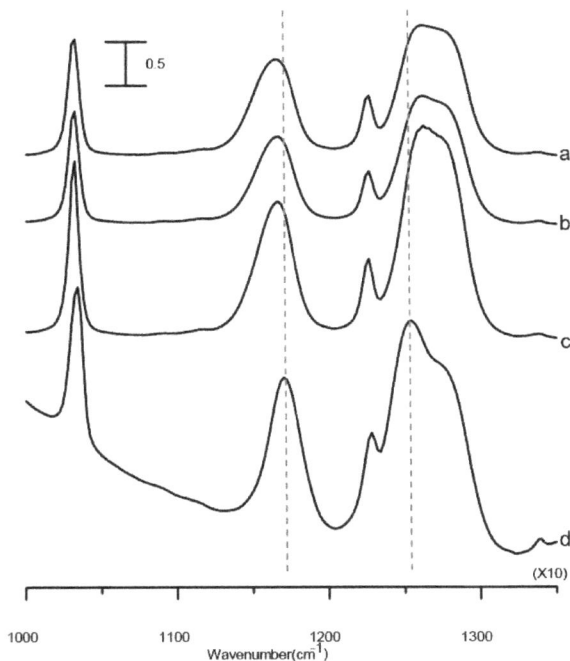

Figure 2. IR spectra of pure [BMIM][TFS] (curve a), the [BMIM][TFS]/Al$_2$O$_3$ mixture (curve b), the mixture after two washes with ethanol (curve c), and the mixture after five washes (curve d), recorded under ambient pressure. Relative intensities of the IR bands are given in increments of 0.05; the scale bar is displayed in the upper left corner of the figure.

Figure 3 displays the IR spectra of a sample from a [BMIM][TFS]/Al$_2$O$_3$ mixture obtained under ambient pressure (curve a) and at 0.4 (curve b), 0.7 (curve c), 1.1 (curve d), 1.5 (curve e), 1.8 (curve f), and 2.5 GPa (curve g). The elevation of pressure to 0.4 GPa leads to a phase transition, and the imidazolium C-H bands are blue-shifted to 3102, 3132, and 3168 cm^{-1} in Figure 3b. The sharper structures of imidazolium C-H bands revealed in Figure 3b–g are in part due to the anisotropic environment in a solid structure. We note that spectral features of the [BMIM][TFS]/Al$_2$O$_3$ mixture in Figure 3 are almost identical to those of pure [BMIM][TFS], as revealed in our previous studies [21]. The results of Figure 3 indicate that the presence of nano-Al$_2$O$_3$ does not perturb ion/ion interactions from the perspective of the C-H stretching motions of the cations. Instead, the blue-shifts revealed in Figure 3 may originate from the combined effect of the overlap repulsion enhanced by hydrostatic pressure, C-H—O contacts, and so forth. The increased understanding of weak interactions, such as hydrogen bonding between the C-H groups on the imidazolium cation and oxygen atoms of the

TFS$^-$ anion, has required an expansion of the classical definition of hydrogen bonding. One of the intriguing aspects of C-H—O interactions is that the C-H bond tends to shorten as a result of the formation of blue-shifting hydrogen bonds. We demonstrate in Figure 3 that high pressure is a valuable method to probe pressure-enhanced blue-shifting of C-H hydrogen bonds in the [BMIM][TFS]/Al$_2$O$_3$ mixture. Scheiner's group and Dannenberg's group indicated that the origin of both red-shifting and blue-shifting hydrogen bonds may be the same, and suggested that both types of hydrogen bonds result from a combination of electrostatic, polarization, charge transfer, dispersion, and exchange/steric repulsion forces between proton donors and acceptors [22,23].

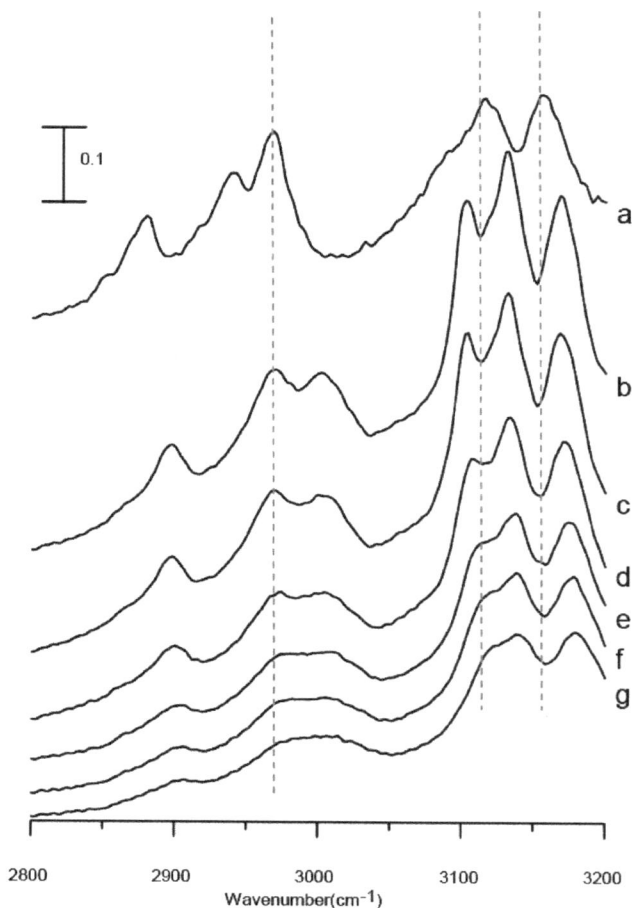

Figure 3. IR spectra of the [BMIM][TFS]/Al$_2$O$_3$ mixture obtained under ambient pressure (curve a) and at 0.4 (curve b), 0.7 (curve c), 1.1 (curve d), 1.5 (curve e), 1.8 (curve f), and 2.5 GPa (curve g). Relative intensities of the IR bands are given in increments of 0.1; the scale bar is displayed in the upper left corner of the figure.

Figure 4 shows the infrared spectra of the [BMIM][TFS]/Al$_2$O$_3$ mixture attributed to the absorption of TFS$^-$ anions under various pressures. As the mixture is compressed to 0.4 GPa in Figure 4b, the 1270 cm^{-1} (doublet) band narrowed and the band at 1165 cm^{-1} splits into two distinct bands at ca. 1152 and 1175 cm^{-1}. These spectral changes may indicate a pressure-induced phase transition arising from both structural reorganization (due to ordered local structures) and a decrease in inhomogeneous broadening. The spectral features of the [BMIM][TFS]/Al$_2$O$_3$ mixture revealed in Figure 4 are almost identical to those of pure [BMIM][TFS]. In contrast to the bandwidth narrowing of asymmetric SO$_3$ stretching vibration at ca. 1270 cm^{-1}, the 1032 cm^{-1} band (symmetric SO$_3$ stretching) displays anomalous bandwidth broadening upon compression in Figure 4b–g. It is likely that transverse optic (TO)-longitudinal optic (LO) splitting of the symmetric SO$_3$ stretching band plays a non-negligible role in Figure 4 [11,12,24].

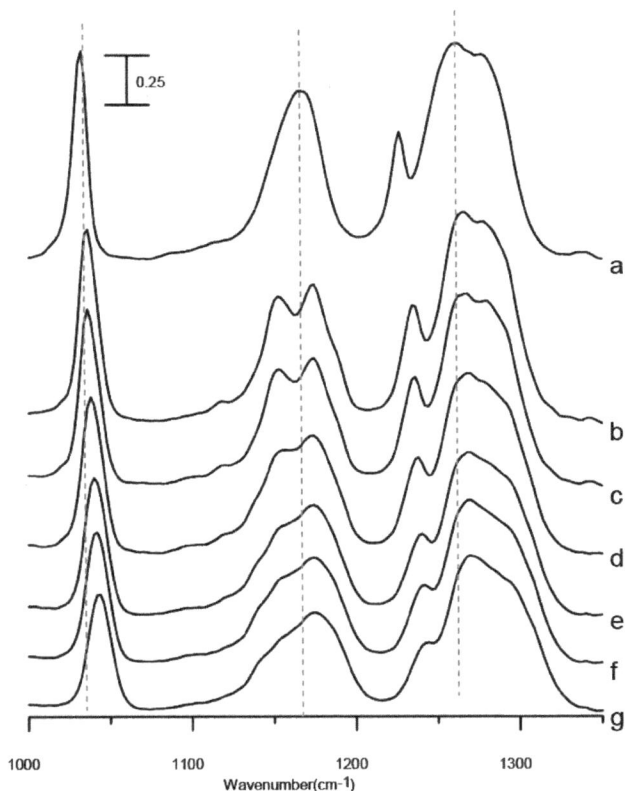

Figure 4. IR spectra of the [BMIM][TFS]/Al$_2$O$_3$ mixture obtained under ambient pressure (curve a) and at 0.4 (curve b), 0.7 (curve c), 1.1 (curve d), 1.5 (curve e), 1.8 (curve f), and 2.5 GPa (curve g). Relative intensities of the IR bands are given in increments of 0.25; the scale bar is displayed in the upper left corner of the figure.

Figure 5 displays the IR spectra of a [BMIM][TFS]/Al$_2$O$_3$ mixture after five washes with ethanol obtained under ambient pressure (curve a) and at 0.4 (curve b), 0.7 (curve c), 1.1 (curve d), 1.5 (curve e), 1.8 (curve f), and 2.5 GPa (curve g). As shown in Figure 5a,b, no appreciable changes in the spectral features of C-H vibrations occur as the sample is compressed to 0.4 GPa. This result is remarkably different from that is revealed for the [BMIM][TFS]/Al$_2$O$_3$ mixture in Figure 3b. It appears that the presence of nano-Al$_2$O$_3$ has an influence on the supramolecular assemblies of the remaining ILs on the surface of Al$_2$O$_3$ after five washes and under high pressures, as revealed in Figure 5. The nano-Al$_2$O$_3$

surfaces may weaken the cation/anion interactions in the interfacial area as a result of the formation of enhanced ionic liquid/particle surface interactions under high pressures. The Al_2O_3/ionic liquid interactions may also affect the hydrogen-bonded network of interfacial [BMIM][TFS] under high pressures. In contrast to the results obtained under ambient pressure (Figure 1), the local structures of both the imidazolium C-H and alkyl C-H groups are disturbed under high pressures for the mixture after five washes, shown in Figure 5, when compared to the pressure dependence of the [BMIM][TFS]/Al_2O_3 mixture (Figure 3). Therefore, high pressures may have the potential to control the order and strength of ionic liquid/surface interactions in this system.

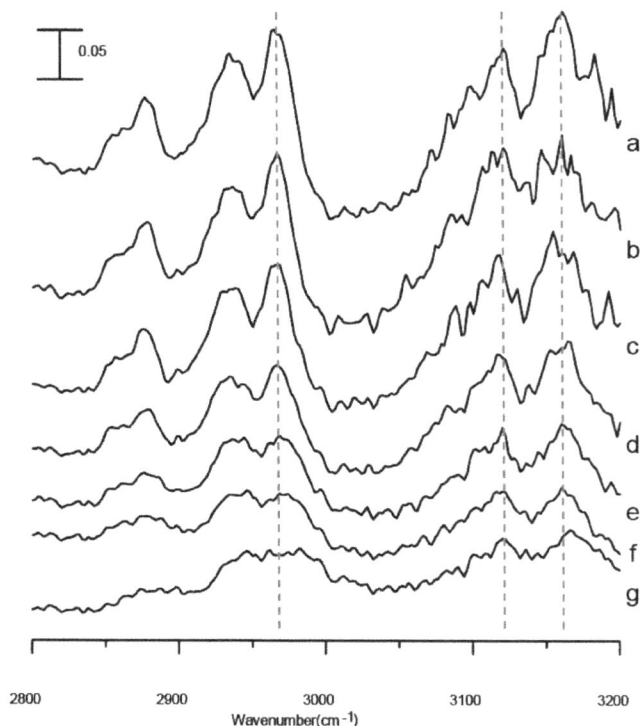

Figure 5. IR spectra of the mixture after five washes with ethanol obtained under ambient pressure (curve a) and at 0.4 (curve b), 0.7 (curve c), 1.1 (curve d), 1.5 (curve e), 1.8 (curve f), and 2.5 GPa (curve g). All curves were baseline-corrected by subtracting straight lines. Relative intensities of the IR bands are given in increments of 0.05; the scale bar is displayed in the upper left corner of the figure.

Figure 6 shows the infrared spectra of the mixture after five washes in the spectral range between 1000 and 1350 cm^{-1} obtained under ambient pressure (curve a) and at 0.4 (curve b), 0.7 (curve c), 1.1 (curve d), 1.5 (curve e), 1.8 (curve f), and 2.5 GPa (curve g). The absence of drastic band-narrowing and splitting in Figure 6 suggests that interfacial ionic liquid/Al_2O_3 interactions of the mixture after five washes may be enhanced by the application of pressure. As revealed in Figure 6g, the absorption bands display mild blue-shifts in frequency as the mixture is compressed to 2.5 GPa.

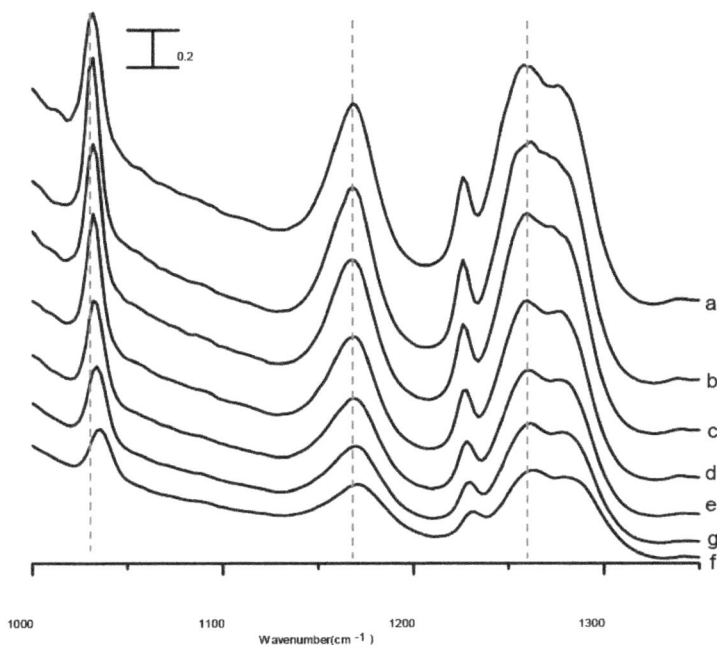

Figure 6. IR spectra of the mixture after five washes with ethanol obtained under ambient pressure (curve a) and at 0.4 (curve b), 0.7 (curve c), 1.1 (curve d), 1.5 (curve e), 1.8 (curve f), and 2.5 GPa (curve g). Relative intensities of the IR bands are given in increments of 0.2; the scale bar is displayed in the upper left corner of the figure.

To illustrate the frequency shift, the pressure dependence of C-H stretching frequencies of the [BMIM][TFS]/Al_2O_3 mixture, the mixture after two washes with ethanol, and the mixture after five washes is plotted in Figure 7. The C-H bands corresponding to the [BMIM][TFS]/Al_2O_3 mixture display significant blue-shifts in frequency as the pressure increases to 0.4 GPa, and then undergo mild shifts in frequency when the pressure increases from 0.4 to 2.5 GPa. The discontinuous jump occurring around 0.4 GPa becomes less obvious for the mixture after two washes with ethanol. In Figure 7, we observe no discontinuous jump in frequency for the mixture after five washes. It is known that the hydrogen bond cooperativity due to concerted charge transfer can greatly enhance the strength of the individual hydrogen bonds involved in the coupling [25]. Hydrogen-bonding non-additivity and the size of clusters are suggested to be responsible for the enhancement of hydrogen bonding [25,26]. Thus, the anomalous jump in frequency at 0.4 GPa can be attributed to both the cooperative and geometric effects of blue-shifting hydrogen bonds. The disappearance of the discontinuous jump upon the reduction of ionic liquid through washing with ethanol may indicate the decrease in strength of blue-shifting hydrogen bonding as C-H/anion interactions are replaced by C-H/Al_2O_3 interactions. The loss of the frequency jump at 0.4 GPa due to washing is related to a reduction of the amount of ionic liquid along the interface. At thicker values, there is likely enough IL to form crystallites. If there is just a monolayer (or about that), crystallization might be difficult and hydrogen bond cooperativity may be reduced.

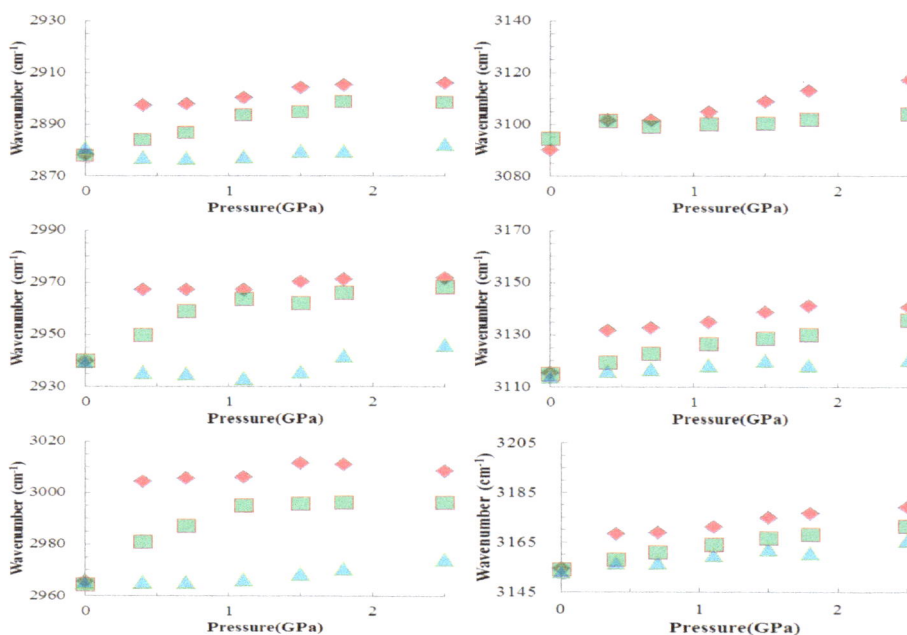

Figure 7. Pressure dependence of the C-H stretching frequencies of the [BMIM][TFS]/Al$_2$O$_3$ mixture (diamonds), the mixture after two washes with ethanol (squares), and the mixture after five washes (triangles).

This study indicates that high pressures can serve as a useful tool to probe the strength of blue-shifting C-H hydrogen bonding in ILs deposited along the surface of nano-Al$_2$O$_3$ surfaces. Furthermore, the surfaces of Al$_2$O$_3$ appear to be capable of breaking or weakening the interactions of the cation/anion clusters, presumably in favor of stable ionic liquid/Al$_2$O$_3$ interactions under high pressures. Supplementary data (IR spectra of the mixture after two washes and pure [BMIM][TFS] with 99% purity) are included in Supplementary Materials (see Figures S1–S3).

4. Conclusions

In this study, we demonstrate that the effect of nano-Al$_2$O$_3$ on hydrogen bonding networks of [BMIM][TFS] can be probed by high-pressure infrared spectroscopy. There are no remarkable changes in the spectral features of [BMIM][TFS] in the presence of nano-Al$_2$O$_3$ at ambient pressure. Nevertheless, nano-Al$_2$O$_3$ may perturb the cation/anion aggregations as a result of the formation of pressure-enhanced ionic liquid/Al$_2$O$_3$ interactions when high pressure is applied to the composite materials. The relative strength of blue-shifting C-H hydrogen bonding, i.e., [BMIM][TFS] > [BMIM]-Al$_2$O$_3$, is also investigated by varying the applied pressure. This study indicates that high pressures may have the potential to tune the strength of ionic liquid/Al$_2$O$_3$ surface interactions. Moreover, high-pressure infrared spectroscopy provides an alternative avenue for a better understanding the interfacial properties of ILs in contact with nanostructured metal oxides.

Supplementary Materials: The following are available online at http://www.mdpi.com/2076-3417/7/8/855/s1, Figure S1: IR spectra of the mixture after two washes with ethanol obtained under ambient pressure (curve a) and at 0.4 (curve b), 0.7 (curve c), 1.1 (curve d), 1.5 (curve e), 1.8 (curve f), and 2.5 GPa (curve g), Figure S2: Infrared spectra of the mixture after two washes with ethanol obtained under ambient pressure (curve a) and at 0.4 (curve b), 0.7 (curve c), 1.1 (curve d), 1.5 (curve e), 1.8 (curve f), and 2.5 GPa (curve g), Figure S3: Infrared

Appl. Sci. **2017**, *7*, 855

spectra of pure [BMIM][TFS] (99% , Lot: N00364.7, UniRegion) obtained under ambient pressure (curve a) and at 0.4 (curve b), 0.7 (curve c), 1.1 (curve d), 1.5 (curve e), 1.8 (curve f), and 2.3 GPa (curve g).

Acknowledgments: The authors thank the National Dong Hwa University and Ministry of Science and Technology (Contract No. MOST 106-2113-M-259-002) of Taiwan for financial support. The authors thank Yen-Hsu Chang and Min-Hsiu Shen for the assistance.

Author Contributions: Hai-Chou Chang is the idea source, writer of the manuscript, and provided data interpretation. Teng-Hui Wang contributed to FTIR spectra measurements. Christopher M. Burba is the co-writer and provided the idea of TO-LO splitting.

Conflicts of Interest: The authors declare no conflict of interest.

References

1. Babucci, M.; Akcay, A.; Balci, V.; Uzan, A. Thermal stability of limits of imidazolium ionic liquids immobilized on metal oxides. *Langmuir* **2015**, *31*, 9163–9176. [CrossRef] [PubMed]
2. Lee, H.J.; Park, J.H. Effect of hydrophobic modification on carbon dioxide absorption using porous alumina (Al$_2$O$_3$) hollow fiber membrane contactor. *J. Membr. Sci.* **2016**, *518*, 79–87. [CrossRef]
3. Pizzozzaro, M.A.; Drobek, M.; Petit, E.; Guerrero, G.; Hesemann, P.; Julbe, A. Design of phosphonated imidazolium-based ionic liquids grafted on γ-alumina: Potential model for hybrid membranes. *Int. J. Mol. Sci.* **2016**, *17*, 1212. [CrossRef] [PubMed]
4. Andanson, J.M.; Baiker, A. Interactions of 1-ethyl-3-methylimidazolium trifluoromethanesulfonate ionic liquid with alumina nanoparticles and organic solvents studied by infrared spectroscopy. *J. Phys. Chem. C* **2013**, *117*, 12210–12217. [CrossRef]
5. Schernich, S.; Kostyshyn, D.; Wagner, V.; Taccardi, N.; Laurin, M.; Wasserscheid, P.; Libuda, J. Interactions between the room-temperature ionic liquid [C$_2$C$_1$Im][OTf] and Pd(111), well-ordered Al$_2$O$_3$, and supported Pd model catalysts from IR spectroscopy. *J. Phys. Chem. C* **2014**, *118*, 3188–3193. [CrossRef]
6. Fuentes, C.E.S.; Guzman-Lucero, D.; Torres-Rodriguez, M.; Likhanoca, N.V.; Bolanos, J.N.; Loivares-Xometl, O.; Lijanova, I.V. CO$_2$/N$_2$ separation using alumina supported membranes based on new functionalized ionic liquids. *Sep. Purif. Technol.* **2017**, *182*, 59–68. [CrossRef]
7. Zhang, S.; Zhang, J.; Zhang, Y.; Deng, Y. Nanoconfined ionic liquids. *Chem. Rev.* **2017**, *117*, 6755–6833. [CrossRef] [PubMed]
8. Singh, D.K.; Rathke, B.; Kiefer, J.; Materny, A. Molecular structure and interactions in the ionic liquid 1-ethyl-3-methylimidazolium trifluoromethanesulfonate. *J. Phys. Chem. A* **2016**, *120*, 6274–6286. [CrossRef] [PubMed]
9. Fedorov, M.V.; Kornyshev, A.A. Ionic liquids at electrified interfaces. *Chem. Rev.* **2014**, *114*, 2978–3036. [CrossRef] [PubMed]
10. Wulf, A.; Fumino, K.; Michalik, D.; Ludwig, R. IR and NMR properties of ionic liquids: Do they tell us the same thing? *ChemPhysChem* **2007**, *8*, 2265–2269. [CrossRef] [PubMed]
11. Burba, C.M.; Janzen, J.; Butson, E.D.; Coltrain, G.L. Using FT-IR spectroscopy to measure charge organization in ionic liquids. *J. Phys. Chem. B* **2013**, *117*, 8814–8820. [CrossRef] [PubMed]
12. Burba, C.M.; Janzen, J.; Butson, E.D.; Coltrain, G.L. Correction to "using FT-IR spectroscopy to measure charge organization in ionic liquids". *J. Phys. Chem. B* **2016**, *120*, 3591–3592. [CrossRef] [PubMed]
13. Chang, H.C.; Hung, T.C.; Wang, H.S.; Chen, T.Y. Local structures of ionic liquids in the presence of gold under high pressures. *AIP Adv.* **2013**, *3*, 032147. [CrossRef]
14. Triolo, A.; Russina, O.; Bleif, H.J. Nanoscale segregation in room temperature ionic liquids. *J. Phys. Chem. B* **2007**, *111*, 4641–4644. [CrossRef] [PubMed]
15. Chang, H.C.; Hsu, D.T. Interactions of ionic liquids and surfaces of graphene related nanoparticles under high pressures. *Phys. Chem. Chem. Phys.* **2017**, *19*, 12269–12275. [CrossRef] [PubMed]
16. Chang, H.C.; Zhang, R.L.; Hsu, D.T. The effect of pressure on cation-cellulose interactions in cellulose/ionic liquid mixtures. *Phys. Chem. Chem. Phys.* **2015**, *17*, 27573–27578. [CrossRef] [PubMed]
17. Chang, H.C.; Jiang, J.C.; Kuo, M.H.; Hsu, D.T.; Lin, S.H. Pressure-enhanced surface interactions between nano-TiO$_2$ and ionic liquid mixtures probed by high pressure IR spectroscopy. *Phys. Chem. Chem. Phys.* **2015**, *17*, 21143–21148. [CrossRef] [PubMed]

Appl. Sci. **2017**, *7*, 855

18. Chang, H.C.; Tsai, T.T.; Kuo, M.H. Using high-pressure infrared spectroscopy to study the interactions between triblock copolymers and ionic liquids. *Macromolecules* **2014**, *47*, 3052–3058. [CrossRef]
19. Wong, P.T.T.; Moffatt, D.J. The uncoupled O-H or O-D stretch in water as an internal pressure gauge for high-pressure infrared spectroscopy of aqueous systems. *Appl. Spectrosc.* **1987**, *41*, 1070–1072. [CrossRef]
20. Wong, P.T.T.; Moffatt, D.J.; Baudias, F.L. Crystalline quartz as an internal pressure calibrant for high-pressure infrared spectroscopy. *Appl. Spectrosc.* **1985**, *39*, 733–735. [CrossRef]
21. Chang, H.C.; Chang, S.C.; Hung, T.C.; Jiang, J.C.; Kuo, J.L.; Lin, S.H. A high-pressure study of the effects of TiO_2 nanoparticles on the structural organization of ionic liquids. *J. Phys. Chem. C* **2011**, *115*, 23778–23783. [CrossRef]
22. Gu, Y.L.; Kar, T.; Scheiner, S. Fundamental properties of the C-H—O interactions: Is it a true hydrogen bond? *J. Am. Chem. Soc.* **1999**, *121*, 9411–9422. [CrossRef]
23. Masunov, A.; Dannenberg, J.J.; Contreras, R.W. C-H bond-shortening upon hydrogen bond formation: Influence of an electric field. *J. Phys. Chem. A* **2001**, *105*, 4737–4740. [CrossRef]
24. Burba, C.M.; Frech, R. Existence of optical phonons in the room temperature ionic liquid 1-ethyl-3-methylimidazolium trifluoromethanesulfonate. *J. Chem. Phys.* **2011**, *134*, 134503. [CrossRef] [PubMed]
25. Knorr, A.; Stange, P.; Fumino, K.; Weinhold, F.; Ludwig, R. Spectroscopic evidence for clusters of like-charged ions in ionic liquids stabilized by cooperative hydrogen bonding. *ChemPhysChem* **2016**, *17*, 458–462.
26. Roth, C.; Appelhagen, A.; Jobst, N.; Ludwig, R. Microheterogeneities in ionic-liquid-methanol solutions studied by FTIR spectroscopy, DFT calculations and molecular dynamics simulations. *ChemPhysChem* **2012**, *13*, 1708–1717. [CrossRef] [PubMed]

applied sciences

MDPI

Article

Nature Inspired Plasmonic Structures: Influence of Structural Characteristics on Sensing Capability

Gerardo Perozziello [1,*], Patrizio Candeloro [1], Maria Laura Coluccio [1], Godind Das [2], Loredana Rocca [1], Salvatore Andrea Pullano [1], Antonino Secondo Fiorillo [1], Mario De Stefano [3] and Enzo Di Fabrizio [1,2]

[1]   Laboratory BioNEM, Department of Experimental and Clinical Medicine, University Magna Graecia of Catanzaro, Loc. Germaneto, 88110 Catanzaro, Italy; Patrizio.candeloro@unicz.it (P.C.); mlcoluccio@gmail.com (M.L.C.); LoredanaRocca94@hotmail.it (L.R.); pullano@unicz.it (S.A.P.); nino@unicz.it (A.S.F.); Enzo.difabrizio@kaust.edu.sa (E.D.F.)
[2]   Physical Sciences and Engineering (PSE), King Abdullah University of Science and Technology (KAUST), Thuwal 23955-6900, Kingdom of Saudi Arabia; gobind.das@kaust.edu.sa
[3]   Environmental Science Department, Second University of Naples, 81100 Caserta, Italy; destefanomario36@gmail.com
*   Correspondence: Gerardo.perozziello@unicz.it; Tel.: +39-393-283-7202

Received: 15 March 2018; Accepted: 23 April 2018; Published: 26 April 2018

Featured Application: The Plasmonic structures presented can be integrated in optical sensors applicable to complex biological mixtures.

Abstract: Surface enhanced Raman scattering (SERS) is a powerful analytical technique that allows the enhancement of a Raman signal in a molecule or molecular assemblies placed in the proximity of nanostructured metallic surfaces, due to plasmonic effects. However, laboratory methods to obtain of these prototypes are time-consuming, expensive and they do not always lead to the desired result. In this work, we analyse structures existing in nature that show, on a nanoscale, characteristic conformations of photonic crystals. We demonstrate that these structures, if covered with gold, change into plasmonic nanostructures and are able to sustain the SERS effect. We study three different structures with this property: opal, a hydrated amorphous form of silica ($SiO_2 \cdot nH_2O$); diatoms, a kind of unicellular alga; and peacock tail feather. Rhodamine 6G (down to 10^{-12} M) is used to evaluate their capability to increase the Raman signal. These results allow us to define an alternative way to obtain a high sensitivity in Raman spectroscopy, currently achieved by a long and expensive technique, and to fabricate inexpensive nanoplasmonic structures which could be integrated into optical sensors.

Keywords: optical sensors; plasmonics; nanostructures

---

1. Introduction

Raman spectroscopy is a powerful vibrational spectroscopy technique that has many advantages. It is an analytical technique with high specificity for the identification of various chemical compounds [1–3]. In addition, it is a non-invasive approach that allows measurements to be performed without any sample preparation, and the measurement time is generally very short. However, the Raman signal is very low and difficult to detect. Statistically, for every 1–10 million photons bombarding a sample, only one will result in Raman scattering [4,5]. For this reason, Raman spectroscopy is often coupled to substrates integrating metal plasmonic nanostructures that reproduce an SERS (surface enhanced Raman scattering) effect [6–8]. The molecules, which need to be analysed, are deposited on these substrates and are excited by an electromagnetic (EM) field, which is enhanced by the proximal metal nanostructures. The amplified EM field interacts with molecules and produces Raman scattering which is again amplified by the metal nanostructures [9–11]. In greater detail, the incident EM waves excite plasmon resonances within the metal nanostructures. Due to resonance effects, the local electric field is enhanced in the proximity of metal nanoparticles. The molecules at the metal surface then experience an enhanced local EM wave, and produce higher Raman signals than those molecules far from the metal surface. Moreover, since Raman signals are a type of EM wave, the same route through plasmon resonances will further enhance the outgoing Raman field. In the end, plasmon resonances are responsible for the double amplification Raman signals of molecules close to metal nanostructures, thus producing a surface-enhanced Raman effect. A detailed review of plasmon-enhanced spectroscopies can be found in [12]. The Raman signal enhancement strongly depends on the size and shape of the metal particles [13–15], and different methods to fabricate them exist in the literature [16–18]. The combination of a top down method, that is, electron beam lithography, with a bottom up technique, that is, chemical electroless deposition, allows a SERS substrate of a self-similar chain of silver nanospheres to be obtained [19]. Plasmonic nanoholes can be made with a template-stripping technique that involves the fabrication of a Si template obtained with electron beam lithography [20]. The Ag-decorated nanotip array, obtained with Electron Beam Induced Deposition and Silver Electroless Deposition, is another efficient 3D plasmonic device, that can be used to obtain a SERS effect [21]. The prototypes described above, although very efficient, are difficult, expensive and time-consuming to fabricate. Other techniques allow the fabrication of nanoplasmonic structures using faster and inexpensive processes. In this regard, metallic colloids [22–24], metallic films [25,26], sensors fabricated by nanosphere lithography [27], and plasmonic systems exploiting nanoparticle self-assembly properties [28,29] have been developed. However, some of the processes to achieve these kinds of low-cost SERS substrates lack reliability in regard to obtaining large sensing surfaces [30,31]. These disadvantages in contrast to the benefits of Raman spectroscopy, which allows information to be obtained easily and without any particular sample preparation [32,33]. Consequently, research involving alternative ways to maintain the SERS effect [34], is of crucial importance. Some structures existing in nature show, on the nanoscale, a conformation called photonic crystal. Photonic crystal, a periodic structure in an optical medium, creates unusual optical dispersion properties, which cause the iridescent colours in some butterfly wings and beetle armours and other structures present in nature [35,36]. These structures, if metallized, can be assimilated to the SERS devices and used as optical sensors [37–39]. Here, we focus on the peacock feather, opal and diatoms, because they are easy to find in nature. We cover their surfaces with gold to change the photonic nanostructure in a plasmonic nanostructure, and we characterize them by detecting Rhodamine, using Raman spectroscopy. Each element shows a different behaviour due to structural characteristics (which we discuss in the following sections). The best response is obtained with opal, allowing the analysis of molecules at very low concentrations and thus confirming their potential application for biosensing devices.

2. Materials and Methods

Metallic nanoparticle aggregates are able to maintain high enhancement of the SERS signal (in the order of 10^{11}) [40]. Experimental evidence shows that by placing two spherical nanoparticles at a distance of less than 5 nm, there is an amplification of the local electric field of several orders of magnitude greater than when the nanoparticles are isolated, at the midpoint between the two. This enhancement of the electric field results from the constructive interference of the surface plasmonic resonances of the two particles. The antiparticle regions with field amplification are called hot spots [41,42]. By metallizing the surface (e.g., with gold), it is possible to assimilate natural nanophotonic structures to metal nanoparticle aggregates, producing plasmonic nanostructures, which can cause high amplification of the SERS signal [43,44].

2.1. Natural Nanophotonic Structures

We used the following natural components:

2.1.1. Peacock Feather

The peacock livery is a typical example of sexual dimorphism. Indeed, the male peacock tail consists of very lengthy feathers that extend to form a 'palette' with the typical eyespot and iridescent colours. This feature is supposed to attract mates and to disorient predators. To investigate the structural properties of the peacock feather, we collected a feather from a male peafowl (*Pavo cristatus*) and focused on its eyespot. At the nanoscale, it is characterized by the presence of cellular organelles, called melanosomes, which are arranged to form a typical two-dimensional crystalline lattice. This arrangement causes the characteristic iridescent colours. The size of the melanosomes vary but on average, they are about 105 nm long, and the distance between the centre of a melanosome and another is about 175 nm. There is a correlation between the diameter of a melanosome and the brightness of cells—brightness increases as the melanosome diameter increases [45].

2.1.2. Opal

Opal is a mineral made of silica, formed by the slow geological deposition of a colloidal gel at a low temperature. Various types of opal exist in nature; the kind of opal we analysed is called *Ethiopian Welo Opal*. It is characterized by its transparency and its vibrant reflections which show all the colours of the rainbow, in particular, yellow, orange, red and green. On the nano-cale, it can be considered a typical example of colloidal crystal, due to its orderly aggregate of particles similar to standard crystal, with subunits of atoms or molecules. The arrangement of amorphous silicon dioxide spheres and their sizes give the structure its optical properties. In fact, the internal colours are determined by the interference and diffraction of light passing through the microstructure of the opal. The spheres are about 10 nm in diameter, while the space between one sphere and another is of a smaller order of magnitude.

2.1.3. Diatoms

Diatoms are unicellular and eukaryotic algae, which can live in isolation or form colonies either in freshwater or marine environments. Diatoms are hundreds of μm in size [46], with a distinctive shell wall consisting of two hydrated silica valves, interconnected in a structure called a frustule. Frustules present patterns of regular arrays of holes, the areolae, which are characterized by sub-micrometric dimensions. There are about 10^5 species of diatoms, whose frustules differ in shape, morphology and size. The dimensions of the frustule pores, typically hundreds of nanometres, provide diatom frustules with peculiar optical properties. In the present work, we considered C. wailesii diatoms, in which organic matter was removed by means of strong acid solutions [47]. The diameter of the valves ranged from 100 to 200 μm, while the average thickness of the wall was about 1 μm. Every single valve is formed by two co-joined plates. The external plate comprises a complex hexagonal arrangement of hollow pores with a diameter ($\cong$200 nm) below the visible wavelength. On the other side, the internal plate of the valve is characterized by hexagonally-spaced pores with diameters in the range of 50 nm and by a lattice constant of about 50 nm.

2.2. Theoretical Simulations

Theoretical simulations of different natural SERS substrates (peafowl, diatoms and opal) were performed using the finite difference time domain method [48]. The designs of these structures were constructed to mimic the actual structures of the devices. The electric field distribution over the nanostructures was calculated using Lumerical, a commercial software package. The mesh size was fixed to 1 nm in the *X*, *Y* and *Z* directions for all calculations. Regarding peafowl, a gold bar (side length = 370 nm, material dispersion from Johnson and Christy [49]) with many overriding functions of dielectric materials (refractive index = 1, side length = 50 nm) was designed in such a manner that the refractive index of the metal area where dielectric bars were inserted became equal to the dielectric one. The gap between the adjacent dielectric materials wais 35 nm. Periodic boundary conditions in the X and Y directions were chosen to calculate the distribution of the electric field (EF). A plane wave with polarization along the horizontal axis at 633 nm was employed to illuminate the peafowl device from the top. Regarding the opal's structure, many gold spheres (radius = 40 nm, material dispersion from Johnson and Christy [49]) were immersed in the dielectric material to resemble the design of an opal. The gap between two gold nanosphere was fixed to 5 nm. Periodic boundary conditions in the X and Y directions were adopted to calculate the distribution of the EF. An incident wavelength of 633 nm was chosen to illuminate the device, and polarization was fixed to an angle of 45°. Regarding the diatoms, we first constructed four circular toroid structures with inner radii of 905 nm and ring widths of 100 nm. The height of the gold ring structures was 92 nm. Within this toroid, many small, circular toroid structures of 105 nm radius (height of gold ring = 10 nm, ring width = 10 nm) were placed randomly. These structures were placed over a dielectric surface (refractive index = 1) deposited over a gold bar. The calculations were made by illuminating the device with an incident wavelength of 633 nm. The incident field considered was 1 V/m.

2.3. SEM Characterization

Several scanning electron microscope (SEM) images of the samples were captured to investigate the sub-micrometre features of the feathers, opals and diatoms, utilizing a Dual Beam scanning electron microscope-focused ion beam (SEM-FIB) Nova 600 NanoLab system (FEI Company, Eindhoven, Netherlands). During the acquisitions, beam energies of 5 and 15 keV, and 45 corresponding electron currents of 0.98 pA and 0.14 nA, were used. Optical images were taken on a Nikon Ti-E, on which a colour camera Nikon DS-Fi2 was mounted (Nikon Instruments SpA, Firenze, Italy).

2.4. Raman Spectroscopy Characterization of the Natural Plasmonic Nanostructures

To determine the effect of SERS on the natural structures, we metallized their surfaces. An apparatus for deposition by sputtering was used to deposit a thin layer of gold (Au) onto the structures. The opal, 10×14.5 mm in size, a sample of a few centimetres cut out from the feather eyespot and diatoms were processed for 30 s, using a current of 50 mA at a voltage of 230 Volts (50 Hz) and a pressure of 100 mbar, to deposit around 10 nm of gold on the samples' surfaces. An Au thickness of 10 nm was chosen to avoid excessive filling of the nanohole structures (this is especially visible in Figure 1C,D). Obviously, the Au thickness was kept constant for all samples to allow a comparison between them. An inVia confocal Raman microscope was used for the optical characterization of the natural plasmonic nanostructures. We set used a wavelength of 633 nm for the laser, a $50\times$ objective and an integration time of 10 s, while the power was set to 2.5 mW for each sample. Spectra were recorded by raster scanning the samples' surfaces, collecting at least 100 spectra, so that for each sample, the point-to-point variation could be estimated. The SERS capabilities of the natural structures were probed using Rhodamine-6G, a fluorescent organic molecule purchased from Sigma Aldrich. A drop of 200 μL of Rhodamine-6G at a concentration of 10^{-5} M was deposited on the opal and feather samples. Meanwhile, 5 mg of diatoms were dissolved in 100 μL of Rhodamine-6G at a concentration of 10^{-5} M and placed on a calcium fluoride slide. Each sample was analysed by recording the Raman signal after the samples had been completely dried in air. Because the opal showed the highest amplification effect (as will be seen in the next sections), 200 μL of Rhodamine-6G at concentrations of 10^{-7} M and 10^{-12} M were deposited on these samples and measurements were recorded. To demonstrate the amplification effect of the mentioned natural structures on Raman spectroscopy, we compared the response of the signal coming from Rhodamine-6G deposited on a flat silicon substrate, covered with gold.

The SERS spectra shown in this work represent the average spectrum of ten spectra recorded at different positions on the samples. Before the comparison between the SERS spectra, a background subtraction was carried out using the spectra recorded on the same samples without R6G (under the same experimental conditions).

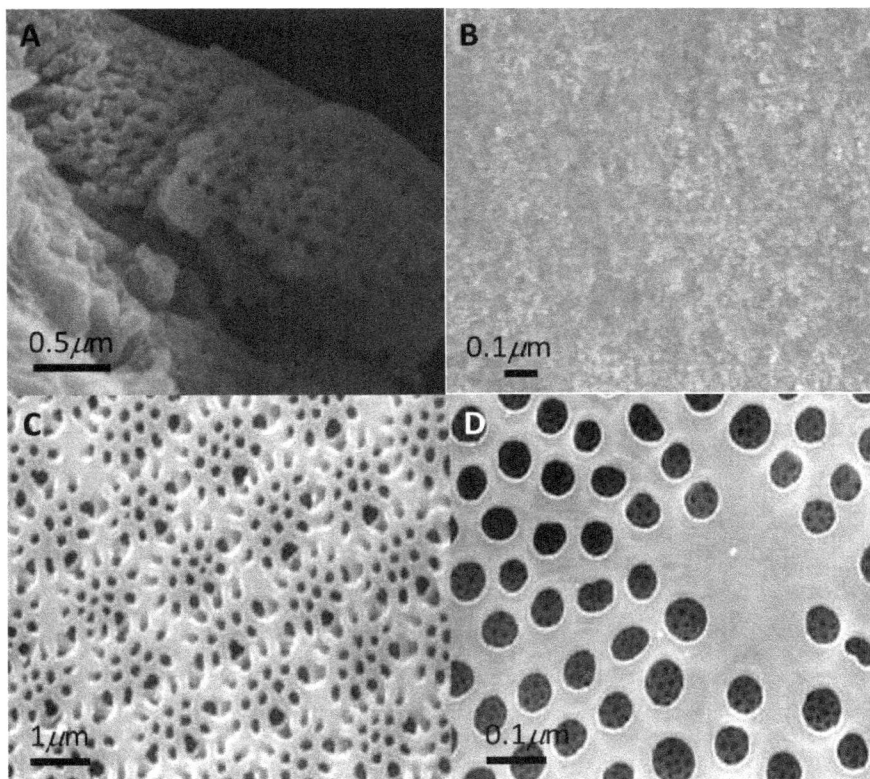

Figure 1. SEM image of the peacock feather (**A**); scanning electron microscope (SEM) images of an opal fragment (**B**); the outer (**C**) and inner (**D**) plates of a single valve of C. wailesii.

3. Results and Discussion

3.1. Optical and SEM Images of the Characterized Samples

Regarding the diatoms, on a micrometric scale it is possible to highlight the sub-micrometric structures randomly placed in the space from regular and repeated perforations in the thickness of the frustules. The diameters of the pores vary, and their dimensions range from 50 nm to 0.5 μm (Figure 1C,D). The opal, on the scanning electron microscope, as shown in Figure 1B, appeared as a homogenous agglomerate of nanospheres with a diameter of tens of nanometres. Taking a SEM image of the peacock feather was very difficult (Figure 1A). In fact, on the micrometre scale, it showed a complex three-dimensional structure. This did not allow us to deposit a thin homogenous layer of gold on it; therefore, SEM imaging was not always clear. However, the SEM image in Figure 1 clearly shows the sub-micrometric structures of the feather.

3.2. Simulated Enhanced Fields

Three different calculations were made for different structures to resemble their associated SERS structures using the FDTD (Finite-Domain Time-Domain) method. The design of the peacock feather is shown in Figure 2A, representing the perspective view of the structure. The calculation shows the confinement of the electric field at the corners of the peafowl structure. The maximum electric field was reached at close to 7 V/m, as shown in Figure 2B. In the case of the opal structure, the localization of electric field in the gap can be observed clearly (Figure 2D). The maximum electric field was 18 V/m. Regarding the diatom structure, non-homogeneous confinement was observed at the edge of the toroid structure and the maximum amplitude was found to be 7 V/m (Figure 2F). The above findings indicate the superior performance of the opal structure as a SERS substrate exited at a wavelength of 633 nm.

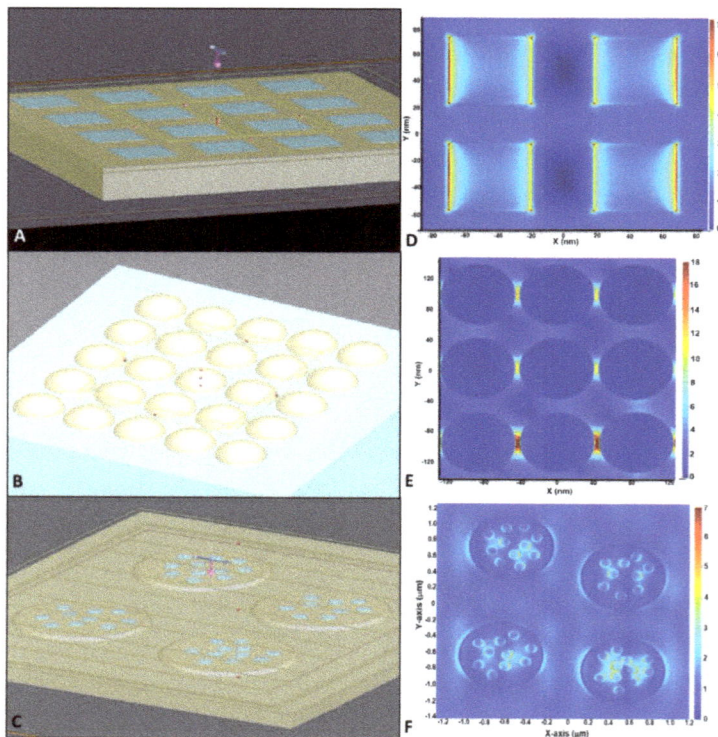

Figure 2. From top left: perspective view of the peacock feather (**A**); opal (**B**) and diatom (**C**) nanostructures. From top right: simulated enhanced electric field of the peacock feather (**D**); opal (**E**) and diatom (**F**) nanostructures.

3.3. Raman Analysis

The preparation of the natural nano structures to be applied in the Raman analysis did not require any particular technique as explained in the Section 2.4. These were used as substrates to perform Raman measurements on samples of Rhodamine 6G. Under the action of the laser excitation, Rhodamine 6G showed peaks at 614, 774, 1182, 1362, 1507 and 1648 cm^{-1}. The 614 cm^{-1} peak corresponds to the C–C–C ring in-plane bending mode. The C–H out-of-plane bending mode for R6G was observed at 774 cm^{-1} while the C–O–C stretching frequency appeared at 1182 cm^{-1}. Peaks centred at 1362, 1507, 1571 and 1648 cm^{-1} were attributed to the aromatic C–C stretch of the R6G molecule [50]. As is evident from the spectra images (Figure 3), typical characteristic peaks from Rhodamine-6G are observed our measurements.

The spectra obtained from each natural structure were processed with the aim of highlighting the qualitative aspects of the results and comparing the measurements. Each sample showed a different behaviour, mainly due to structural reasons. The best response came from the opal (marked in red in the image above). This was due to the small size of its nanospheres, whose diameters are a few dozens of nm, as well as its regular and flat conformation, which allowed, in addition to complete gold coverage, direct interaction between the source laser and the molecule. The spectrum obtained using the peacock feather as support for the Raman analysis, showed visible, but low intensity, peaks (marked in green in the image above). In fact, the irregular structure of the peacock feather does not allow complete gold coverage, and consequently part of the sample was not able to produce the SERS effect. Moreover, since the surface of the feather, at the microscale, was not flat, the deposited solution containing Rhodamine did not stand in position, falling out of the sample in a few minutes. Another limitation was the usable power intensity as high powers can damage the feather. The detected peaks concerning the diatoms (marked in blue in the image above) were indistinguishable from baseline and did not allow the characterization of the molecule. This happened because, while in the peacock feather and the opal, the nanostructures were in direct contact with the analyte and largely exposed to the laser beam, the diatoms 3D tubular structures were casually arranged in the space. In addition, the 3D structure caused the absorption of a portion of the molecule; thus, most of the signal came from diatoms and not from Rhodamine, and that is why the silicon pattern (marked in black in the image above), even without nanostructures, showed peaks higher than those of the diatoms. Furthermore, the irregularity of the diatoms does not allow total gold coverage; the size of the diatom nanostructure is bigger than the nanostructure of the other samples, causing a decrease in the enhancement obtainable. Improvements in the SERS signals were computed as intensity ratios between the natural nanostructures and flat Si–Au samples for the peak at 1358 cm^{-1}. We found that the SERS improvement for opal was approximately 13.88, peafowl had a much lower value of 1.88, while for diatom, the value of 0.63 indicated that the Si–Au substrate behaved much better as SERS substrate. Considering that SERS intensities scale like the fourth power of the electric field, the just-mentioned ratios correspond to the following improvements in the enhancing factors: 1.93 for opal, 1.17 for peafowl, and 0.89 for diatom. It is worth remarking that these values are not the absolute enhancing factors of the electric field, but only the relative values with reference to Si–Au SERS enhancing factor. As already reported the literature (see as an example [51]), the overall SERS intensity in plasmonic nanostructures is strongly dominated by only few positions where the highest electric field enhancement is achieved (hot spots). By comparing the signal intensities at 1361 cm^{-1}, we observed a SERS ratio of approximately 3 between the opal and the peacock feather samples. On the other hand, by comparing the maximum enhancement factors of the electric fields obtained with the FDTD simulations, the same SERS ratio was approximately 36 between the opal and peacock samples (taking into account that SERS signal scale up like the fourth power of the electric field). However, it should be considered that the spatial pitch (1D) of the opal structures was double that of the pitch of the peacock feather, and consequently the opal sample had a number of hot spots per surface unit that were four times smaller than those of the peacock feather. As a conclusion, the FDTD simulations provided an SERS ratio of approximately 9 between the opal and peacock samples. The further disagreement between the FDTD and experimental values was likely due to the structural variations intrinsically expected in natural nanostructures. Finally, even though the maximum enhancement factor of the electric field in the diatom hot spots was similar to that of peacock feather, the number of hot spots per unit surface was much smaller in diatoms, thus explaining the low SERS intensity observed in the experiments. In conclusion, the positive performance shown by the opal allowed a greater decrease in the Rhodamine-6G concentration. Using this sample, Rhodamine-6G was detected at a value of 10^{-12} M (Figure 4).

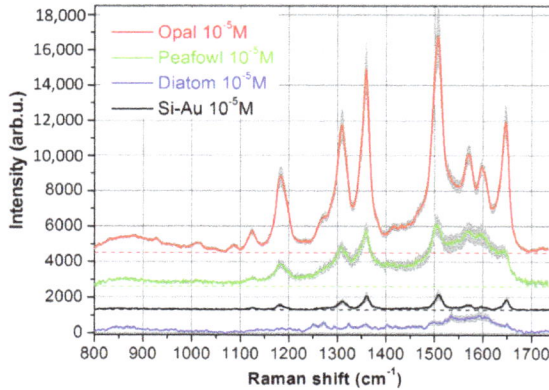

Figure 3. Raman spectra of Rhodamine-6G dispersed in a solution at a concentration of 10^{-5} M measured on opal fragments (opal), peacock feathers (peafowl), diatoms (diatom) and flat silicon surface (Si–Au). It is possible to note the characteristic peaks of the Rhodamine at 1182, 1361, 1508 and 1647 cm^{-1}. The gray shadow of each curve displays the standard deviations of the SERS spectra.

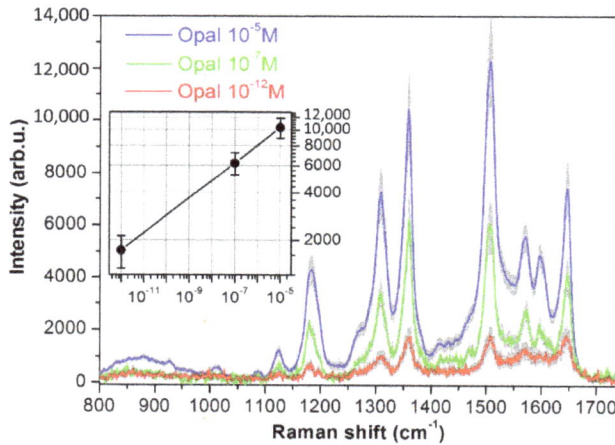

Figure 4. Raman spectra of Rhodamine-6G measured at a different concentration on opal fragments. The grayshadow of each curve displays the standard deviations of the SERS spectra. In the inset, the 1361 cm^{-1} peak intensity is reported as a function of the R6G concentration, in a log–log plot.

4. Conclusions

Nowadays, the fabrication of prototypes able to sustain the SERS effect requires high costs and time-consuming procedures, which do not always allow the expected result to be achieved. In this work, we exploited samples existing in nature to obtain SERS devices capable of increasing the Raman signal. The integration of natural structures into Raman analysis is linked to a lower cost, speed and greater ease of use. Each analysed sample showed a different response due to structural reasons. The diatoms were not suitable for the experiment, the peacock feather could be suitable with some modifications and improvements, and the opal was an efficient medium that was able to determine a positive response even at very low concentrations. This result could become a relevant device for the Raman spectroscopy community.

Author Contributions: E.D.F., A.S.F. and G.P. conceived and designed the experiments; L.R., S.A.P. and M.L.C. performed the experiments; G.D. performed the simulations; G.D. and P.C. analyzed the data; M.D.S. contributed reagents/materials/analysis tools; G.P., L.R. and G.D. wrote the paper.

Acknowledgments: This work was supported by the project for Young researchers financed from the Italian Ministry of Health "High Throughput analysis of cancer cells for therapy evaluation by microfluidic platforms integrating plasmonic nanodevices" (CUP J65C13001350001, project No. GR-2010-2311677) granted to the nanotechnology laboratory of the Department of Experimental and Clinical Medicine of the University "Magna Graecia" of Catanzaro.

Conflicts of Interest: The authors declare no conflict of interest.

References

1. DeVetter, B.M.; Mukherjee, P.; Murphy, C.J.; Bhargava, R. Measuring binding kinetics of aromatic thiolated molecules with nanoparticles via surface-enhanced Raman spectroscopy. *Nanoscale* **2015**, *7*, 8766–8775. [CrossRef] [PubMed]

2. Geng, J.; Aioub, M.; El Sayed, M.A.; Barry, B.A. An Ultraviolet Resonance Raman Spectroscopic Study of Cisplatin and Transplatin Interactions with Genomic DNA. *J. Phys. Chem. B* **2017**, *121*, 8975–8983. [CrossRef] [PubMed]

3. Vo-Dinh, T.; Fales, A.M.; Griffin, G.D.; Khoury, C.G.; Liu, Y.; Ngo, H.; Norton, S.J.; Register, J.K.; Wang, H.N.; Yuan, H. Plasmonic nanoprobes: From chemical sensing to medical diagnostics and therapy. *Nanoscale* **2013**, *5*, 10127–10140. [CrossRef] [PubMed]

4. Zavaleta, C.L.; Kircher, M.F.; Gambhir, S.S. Raman's "Effect" on Molecular Imaging. *J. Nucl. Med.* **2011**, *52*, 1839–1844. [CrossRef] [PubMed]

5. Le Ru, E.; Etchegoin, P.G. *Principles of Surface-Enhanced Raman Spectroscopy*; Elsevier: New York, NY, USA, 2008; ISBN 978-0-444-52779-0.

6. Kim, D.; Campos, A.R.; Datt, A.; Gao, Z.; Rycenga, M.; Burrows, N.D.; Greeneltch, N.G.; Murphy, C.A.M.C.J.; van Duyne, R.P.; Haynes, C.L. Microfluidic-SERS Devices for One Shot Limit-of-Detection. *Analyst* **2014**, *139*, 3227–3234. [CrossRef] [PubMed]

7. Hamon, C.; Liz-Marzán, L.M. Colloidal Design of Plasmonic Sensors Based on Surface Enhanced Raman Scattering. *J. Colloid Interface Sci.* **2018**, *512*, 834–843. [CrossRef] [PubMed]

8. Gentile, F.; Das, G.; Coluccio, M.L.; Mecarini, F.; Accardo, A.; Tirinato, L.; Tallerico, R.; Cojoc, G.; Liberale, C.; Candeloro, P.; et al. Ultra low concentrated molecular detection using super hydrophobic surface based biophotonic devices. *Microelectron. Eng.* **2010**, *87*, 798–801. [CrossRef]

9. Das, G.; Coluccio, M.L.; Alrasheed, S.; Giugni, A.; Allione, M.; Torre, B.; Perozziello, G.; Candeloro, P.; Di Fabrizio, E. Plasmonic nanostructures for the ultrasensitive detection of biomolecules. *Riv. Nuovo Cimento* **2016**, *39*, 547–586.

10. Perozziello, G.; Giugni, A.; Allione, M.; Torre, B.; Das, G.; Coluccio, M.L.; Marini, M.; Tirinato, L.; Moretti, M.; Limongi, T.; et al. Nanoplasmonic and microfluidic devices for biological sensing. In *NATO Science for Peace and Security Series B: Physics and Biophysics*; Springer: Berlin, Germany, 2017; pp. 247–274.

11. Perozziello, G.; Candeloro, P.; de Grazia, A.; Esposito, F.; Allione, M.; Coluccio, M.L.; Tallerico, R.; Valpapuram, I.; Tirinato, L.; Das, G.; et al. Microfluidic device for continuous single cells analysis via Raman spectroscopy enhanced by integrated plasmonic nanodimers. *Opt. Express* **2016**, *24*, A180–A190. [CrossRef] [PubMed]

12. Itoh, T.; Yamamoto, Y.S.; Ozaki, Y. Plasmon-enhanced spectroscopy of absorption and spontaneous emissions explained using cavity quantum optics. *Chem. Soc. Rev.* **2017**, *46*, 3904–3921. [CrossRef] [PubMed]

13. Gao, Z.; Burrows, N.D.; Valley, N.A.; Schatz, G.C.; Murphy, C.J.; Haynes, C.L. In Solution SERS Sensing Using Mesoporous Silica-Coated Gold Nanorods. *Analyst* **2016**, *141*, 5088–5095. [CrossRef] [PubMed]

14. Hooshmand, N.; Mousavi, H.S.; Panikkanvalappil, S.R.; Adibi, A.; El-Sayed, M.A. High-sensitivity molecular sensing using plasmonic nanocube chains in classical and quantum coupling regimes. *Nano Today* **2017**, *17*, 14–22. [CrossRef]

15. Kanipe, K.N.; Chidester, P.P.F.; Stucky, G.D.; Meinhart, C.D.; Moskovits, M. Properly Structured, Any Metal Can Produce Intense Surface Enhanced Raman Spectra. *J. Phys. Chem. C* **2017**, *121*, 14269–14273. [CrossRef]

16. Coluccio, M.L.; Gentile, F.; Das, G.; Nicastri, A.; Perri, A.M.; Candeloro, P.; Perozziello, G.; Proietti, R.; Totero-Gongora, J.S.; Alrasheed, S.; et al. Detection of single amino acid mutation from human breast cancer by plasmonic self-similar chain. *Sci. Adv.* **2015**, *1*, e1500487. [CrossRef] [PubMed]

17. Perozziello, G.; Candeloro, P.; Gentile, F.; Nicastri, A.; Perri, A.M.; Coluccio, M.L.; Parrotta, E.; de Grazia, A.; Tallerico, M.; Pardeo, F.; et al. A microfluidic dialysis device for complex biological mixture SERS analysis. *Microelectron. Eng.* **2015**, *144*, 37–41. [CrossRef]

18. Perozziello, G.; Candeloro, P.; Gentile, F.; Nicastri, A.; Perri, A.; Coluccio, M.L.; Adamo, A.; Pardeo, F.; Catalano, R.; Parrotta, E.; et al. Microfluidics & Nanotechnology: Towards fully integrated analytical devices for the detection of cancer biomarkers. *RSC Adv.* **2014**, *4*, 55590–55598.

19. Coluccio, M.L.; Gentile, F.; Francardi, M.; Perozziello, G.; Malara, N.; Candeloro, P.; Di Fabrizio, E. Electroless Deposition and Nanolithography Can Control the Formation of Materials at the Nano-Scale for Plasmonic Applications. *Sensors* **2014**, *14*, 6056–6083. [CrossRef] [PubMed]

20. Candeloro, P.; Iuele, E.; Perozziello, G.; Coluccio, M.L.; Gentile, F.; Malara, N.; Mollace, V.; Di Fabrizio, E. Plasmonic nanoholes as SERS devices for biosensing applications: An easy route for nanostructures fabrication on glass substrates. *Microelectron. Eng.* **2017**, *175*, 30–33. [CrossRef]

21. Coluccio, M.L.; Francardi, M.; Gentile, F.; Candeloro, P.; Ferrara, L.; Perozziello, G.; Di Fabrizio, E. Plasmonic 3D-structures based on silver decorated nanotips for biological sensing. *Opt. Lasers Eng.* **2016**, *76*, 45–51. [CrossRef]

22. Vinod, M.; Gopchandran, K.G. Au, Ag and Au:Ag colloidal nanoparticles synthesized by pulsed laser ablation as SERS substrates. *Prog. Nat. Sci. Mater. Int.* **2014**, *24*, 569–578. [CrossRef]

23. Cyrankiewicz, M.; Wybranowski, T.; Kruszewski, S. Study of SERS efficiency of metallic colloidal systems. *J. Phys. Conf. Ser.* **2007**, *79*, 012013. [CrossRef]

24. Kneipp, K.; Kneipp, H.; Manoharan, R.; Hanlon, E.B.; Itzkan, I.; Dasari, R.R.; Feld, M.S. Extremely Large Enhancement Factors in Surface-Enhanced Raman Scattering for Molecules on Colloidal Gold Clusters. *Appl. Spectrosc.* **1998**, *52*, 1493–1497. [CrossRef]

25. Lee, C.; Robertson, C.S.; Nguyen, A.H.; Kahraman, M.; Wachsmann-Hogiu, S. Thickness of a metallic film, in addition to its roughness, plays a significant role in SERS activity. *Sci Rep.* **2015**, *5*, 11644. [CrossRef] [PubMed]

26. Botta, R.; Rajanikanth, A.; Bansal, C. Silver nanocluster films for glucose sensing by Surface Enhanced Raman Scattering (SERS). *Sens. Bio-Sens. Res.* **2016**, *9*, 13–16. [CrossRef]

27. Pisco, M.; Galeotti, F.; Quero, G.; Grisci, G.; Micco, A.; Mercaldo, L.V.; Veneri, P.D.; Cutolo, A.; Cusano, A. Nanosphere lithography for optical fiber tip nanoprobes. *Light Sci. Appl.* **2017**, *6*, e16229. [CrossRef]

28. Garoff, S.; Weitz, D.A.; Alvarez, M.S.; Chung, J.C. Electromagnetically Induced Changes in Intensities, Spectra and Temporal Behavior of Light Scattering from Molecules on Silver Island Films. *J. Phys. Colloq.* **1983**, *44*, C10-345–C10-348. [CrossRef]

29. Edel, J.B.; Kornyshev, A.A.; Urbakh, M. Self-Assembly of Nanoparticle Arrays for Use as Mirrors, Sensors and Antennas. *ACS Nano* **2013**, *7*, 9526–9532. [CrossRef] [PubMed]

30. Zhang, X.; Young, M.A.; Lyandres, O.; van Duyne, R.P. Rapid detection of an anthrax biomarker by surface-enhanced Raman spectroscopy. *J. Am. Chem. Soc.* **2005**, *127*, 4484–4489. [CrossRef] [PubMed]

31. Cho, H.; Kumar, S.; Yang, D.; Vaidyanathan, S.R.; Woo, K.; Garcia, I.; Shue, H.J.; Yoon, Y.; Ferreri, K.; Choo, H. SERS-Based Label-Free Insulin Detection at Physiological Concentrations for Analysis of Islet Performance. *ACS Sens.* **2018**, *3*, 65–71. [CrossRef] [PubMed]

32. Ngo, H.T.; Wang, H.; Burke, T.; Ginsburg, G.S.; Vo-Dinh, T. Multiplex detection of disease biomarkers using SERS molecular sentinel-on-chip. *Anal. Bioanal. Chem.* **2014**, *406*, 3335–3344. [CrossRef] [PubMed]

33. Ngo, H.T.; Wang, H.; Fales, A.M.; Vo-Dinh, T. Plasmonic SERS biosensing nanochips for DNA detection. *Anal. Bioanal. Chem.* **2016**, *408*, 1773–1781. [CrossRef] [PubMed]

34. Kanipe, K.N.; Chidester, P.P.F.; Stucky, G.D.; Moskovits, M. Large Format Surface-Enhanced Raman Spectroscopy Substrate Optimized for Enhancement and Uniformity. *ACS Nano* **2016**, *10*, 7566–7571. [CrossRef] [PubMed]

35. Srinivasarao, M. Nano-Optics in the biological World: Beetles, Butterflies, Birds, and Moths. *Chem. Rev.* **1999**, *99*, 1935–1961. [CrossRef] [PubMed]

36. Wu, L.Y.; Ross, B.M.; Hong, S.; Lee, L.P. Bioinspired nanocorals with decoupled cellular targeting and sensing functionality. *Small* **2010**, *6*, 503–507. [CrossRef] [PubMed]

37. Zhang, Q.X.; Chen, Y.X.; Guo, Z.; Liu, H.L.; Wang, D.P.; Huang, X.J. Bioinspired multifunctional hetero-hierarchical micro/nanostructure tetragonal array with self-cleaning, anticorrosion, and concentrators for the SERS detection. *ACS Appl. Mater. Interfaces* **2013**, *5*, 10633–10642. [CrossRef] [PubMed]

38. Ren, F.; Campbell, J.; Hasan, D.; Wang, X.; Rorrer, G.L.; Wang, A.X. Bio-inspired plasmonic sensors by diatom frustules. In Proceedings of the Lasers and Electro-Optics (CLEO), San Jose, CA, USA, 9–14 June 2013; IEEE: Piscataway, NJ, USA, 2013; pp. 1–2.

39. Hong, S.; Lee, M.Y.; Jackson, A.O.; Lee, P.L. Bioinspired optical antennas: Gold plant viruses. *Light Sci. Appl.* **2015**, *4*, e267. [CrossRef]

40. Gentile, F.; Coluccio, M.L.; Accardo, A.; Marinaro, G.; Rondanina, E.; Santoriello, S.; Marra, S.; Das, G.; Tirinato, L.; Perozziello, G.; et al. Tailored Ag nanoparticles/nanoporous superhydrophobic surfaces hybrid devices for the detection of single molecule. *Microelectron. Eng.* **2012**, *97*, 349–352. [CrossRef]

41. Li, W.; Camargo, P.H.C.; Lu, X.; Xia, Y. Dimers of Silver Nanospheres: Facile Synthesis and Their Use as Hot Spots for Surface-Enhanced Raman Scattering. *Nano Lett.* **2009**, *9*, 485–490. [CrossRef] [PubMed]

42. Velleman, L.; Scarabelli, L.; Sikdar, D.; Kornyshev, A.A.; Liz-Marzán, L.M.; Edel, J.B. Monitoring Plasmon Coupling and SERS Enhancement Through in situ Nanoparticle Spacing Modulation. *Faraday Discuss* **2017**, *205*, 67–83. [CrossRef] [PubMed]

43. Geldmeier, J.A.; Mahmoud, M.A.; Jeon, J.W.; El-Sayed, M.; Tsukruk, V.V. The effect of plasmon resonance coupling in P3HT-coated silver nanodisk monolayers on their optical sensitivity. *J. Mater. Chem. C* **2016**, *4*, 9813–9822. [CrossRef]

44. Bosnick, K.A.; Wang, H.M.; Haslett, T.L.; Moskovits, M. Quantitative Determination of the Raman Enhancement of $Ag_{30}(CO)_{25}$ and $Ag_{50}(CO)_{40}$ Matrix Isolated in Solid Carbon Monoxide. *J. Mater. Chem. C* **2016**, *120*, 20506–20511. [CrossRef]

45. Han, J.; Su, H.; Song, F.; Gu, J.; Zhang, D.; Jiang, L. Novel Photonic Crystals: Incorporation of Nano-CdS into the Natural Photonic Crystals within Peacock Feathers. *Langmuir* **2009**, *25*, 3207–3211. [CrossRef] [PubMed]

46. De Tommasi, E.; Rea, I.; Mocella, V.; Moretti, L.; De Stefano, M.; Rendina, I.; De Stefano, L. Multi-wavelength study of light transmitted through a single marine centric diatom. *Opt. Express* **2010**, *18*, 12203–112212. [CrossRef] [PubMed]

47. De Stefano, M.; De Stefano, L. Nanostructures in diatom frustules: Functional morphology of valvocopulae in cocconeidacean monoraphid Taxa. *J. Nanosci. Nanotechnol.* **2005**, *5*, 1–10. [CrossRef]

48. Yu, R.; Liz-Marzán, L.M.; García de Abajo, F.J. Universal Analytical Modeling of Plasmonic Nanoparticles. *Chem. Soc. Rev.* **2017**, *46*, 6710–6724. [CrossRef] [PubMed]

49. Johnson, P.B.; Christy, R.W. Optical Constants of the Noble Metals. *Phys. Rev. B* **1972**, *6*, 4370. [CrossRef]

50. Sil, S.; Kuhar, N.; Acharya, S.; Umapathy, S. Is Chemically Synthesized Graphene'Really' a Unique Substrate for SERS and Fluorescence Quenching? *Sci. Rep.* **2013**, *3*, 3336. [CrossRef] [PubMed]

51. Yoshida, K.; Itoh, T.; Tamaru, H.; Biju, V.; Ishikawa, M.; Ozaki, Y. Quantitative evaluation of electromagnetic enhancement in surface-enhanced resonance Raman scattering from plasmonic properties and morphologies of individual Ag nanostructures. *Phys. Rev. B* **2010**, *81*, 115406. [CrossRef]

MDPI

St. Alban-Anlage 66

4052 Basel

Switzerland

Tel. +41 61 683 77 34

Fax +41 61 302 89 18

www.mdpi.com

Applied Sciences Editorial Office

E-mail: applsci@mdpi.com

www.mdpi.com/journal/applsci

www.ingramcontent.com/pod-product-compliance
Lightning Source LLC
Chambersburg PA
CBHW051912210326
41597CB00033B/6121